Project Management in Extreme Situations

Lessons from Polar Expeditions, Military and Rescue Operations, and Wilderness Explorations

Project Management in Extreme Situations

Lessons from Polar Expeditions, Military and Rescue Operations, and Wilderness Explorations

Edited by
Monique Aubry
Pascal Lièvre

CRC Press is an imprint of the
Taylor & Francis Group, an **informa** business
AN AUERBACH BOOK

CRC Press
Taylor & Francis Group
6000 Broken Sound Parkway NW, Suite 300
Boca Raton, FL 33487-2742

First issued in paperback 2022

ISBN 13: 978-1-03-247705-3 (pbk)
ISBN 13: 978-1-4822-0882-5 (hbk)

DOI: 10.1201/9781315373928

Publisher's Note
The publisher has gone to great lengths to ensure the quality of this reprint but points out that some imperfections in the original copies may be apparent.

Visit the Taylor & Francis Web site at
http://www.taylorandfrancis.com

and the CRC Press Web site at
http://www.crcpress.com

Dedication

To Jacques LeBlanc, my late father-in-law, who inspired me by his lifetime research on the *Man in the Cold*.*

To our families, who have continually given us their love and still do.

– Monique Aubry
– Pascal Lièvre

* **LeBlanc, J. (1975).** *Man in the Cold*. Springfield, Ill: Charles C. Thomas.

Contents

Dedication v

Contents vii

Foreword xvii
Ann Langley

Preface xxi

Acknowledgments xxvii

About the Authors xxix

Monique Aubry, David Brazeau, Christophe N. Bredillet, Maude Brunet, Anne-Marie Cabana, Danielle Desbiens, Gilles Garel, Anaïs Gautier, Frédéric Gautier, Cécile Godé, Alain A. Grenier, Nathalie Guérard, Markus Hällgren, Benoit Lalonde, Ann Langley, Marc Lecoutre, Valérie Lehmann, Pascal Lièvre, Jean Martel, Tessa Melkonian, Thierry Picq, Jean-Pierre Polonovski, Michel Récopé, Géraldine Rix-Lièvre, Linda Rouleau

Introduction: Blowing Hot and Cold on Project Management xxxix
Christophe N. Bredillet

***Part One* – Polar Expeditions** 1

Chapter 1 A Polar Expedition Project and Project Management 3
Gilles Garel & Pascal Lièvre

1.1. The Expedition Project in the Light of Bruno Latour's Thought 5

1.1.1. Actors and Objects 5
1.1.2. Controversies 7
1.2. The Expedition Experience and Insights for Project Management 8
1.2.1. Team Formation: Prototyping Potential Situations and Managing Expectations 8
1.2.2. The Relationship between Preparation and Action: Adapting and Preparing to Improvise, Rather than Planning 11
1.2.3. Knowledge and Ignorance: Knowledge in Action 12
1.3. Conclusion 14
References 15

Chapter 2 Ambidexterity as a Project Leader Competency: A Comparative Case Study of Two Polar Expeditions
Monique Aubry & Pascal Lièvre **17**

2.1. Suggested Analytical Framework 18
2.1.1. The Dual Nature of Project Ambidexterity 18
2.1.2. Ambidexterity According to March 19
2.1.3. Ambidexterity According to Mintzberg 20
2.2. Comparative Case Study of Two Polar Expeditions 21
2.2.1. Methodology 21
2.2.2. The Two Expeditions 23
2.3. Management Methods Used in a Critical Situation during Each Expedition 24
2.3.1. The ARC Expedition's Critical Situation: Protecting against Bear Attacks 25
2.3.2. The ANT Expedition's Critical Situation: Mooring the Ship in the Bay of Whales 26
2.4. Conclusion 27
References 29

Chapter 3 Mobilization and Sensibility on Polar Expeditions: More than Mere Motivation **31**
Michel Récopé, Pascal Lièvre, & Géraldine Rix-Lièvre

3.1. Contributions from Motivational Psychology and the Philosophy of Experience to the Notion of Commitment 32
3.1.1. Motivational Psychology 32
3.1.2. The Philosophy of Experience 34

3.2. Expressed Motivation and Mobilization during a Polar Expedition 37
3.2.1. Polar Discovery as a Common Motivation 37
3.2.2. Observable Evidence of Mobilization 38
3.2.3. The Financial Investment 40
3.2.4. The Issue of Sensibility 40
3.2.5. Sensibility Differences and Expedition Conduct 41
3.3. Conclusion 42
References 43

Chapter 4 Mobilizing Social Networks beyond Project Team Boundaries: The Case of Polar Expeditions 45
Marc Lecoutre & Pascal Lièvre

4.1. Two Case Studies on How Social Networks Are Mobilized for the Organization of Polar Expeditions 47
4.1.1. Joël's Polar Expedition 47
4.1.2. Luc's Polar Expedition 48
4.2. Cooperation in Project Management and the Types of Ties Mobilized 49
4.2.1. Advantages and Limitations of the Two Network Mobilization Approaches 49
4.2.2. Cooperation: Strong Tie or Weak Tie? 50
4.3. Three Illustrations of Weak Ties and Cooperative Potential 53
4.3.1. A Weak Tie without Cooperative Potential 53
4.3.2. A Weak Tie Potentially Cooperative from a Utilitarian (or Complementary) Stance 53
4.3.3. A Weak Tie Potentially Cooperative from a Shared-Identity (or Communautarian) Stance 54
4.4. Conclusions: Perspectives on Entry into Cooperation in Project Management 54
References 56

Chapter 5 A Methodology for Investigating the "Actual" Course of a Project: The Case of a Polar Expedition 59
Géraldine Rix-Lièvre & Pascal Lièvre

5.1. The Obstacles to Overcome when Investigating Organizing as It Is Occurring 60

5.1.1. The Limits of Observation 60
5.1.2. From the Limits of Spontaneous Discourse when Studying Activities to a New Context of Expression 61
5.2. The Observatory 63
5.2.1. Considering a Collectively Shared Reality: The Multimedia Logbook 63
5.2.2. Entering Each Actor's World of Practice: The Situated Practices Objectifying System 65
5.2.3. How Should the Two Research Roles Complement Each Other to Better Understand Organizing *in Situ*? 68
5.3. Conclusion 69
References 70

Chapter 6 A Traditional Cree Expedition on the Ancestral Lands of the Neeposh Family of Northern Québec 73
Nathalie Guérard & Anne-Marie Cabana

6.1. Project Organization and Logistics 75
6.2. The Trip and Camps 77
6.2.1. Welcome Feast 77
6.2.2. Setup of a Typical Campsite 78
6.2.3. A Typical Day 79
6.3. Assessment of the Journey 85

Chapter 7 Borrowing Concepts from Expedition Travel to Stimulate Alternative Tourism 87
Alain A. Grenier

7.1. Tourism and the Environment 88
7.2. From Tourism to Adventure 91
7.3. From Adventure to Expedition 95
7.4. The Role of Expeditions in Modern Society 101
7.4.1. Physical Motivation 101
7.4.2. Mental Motivation 102
7.4.3. Identity Motivation 103
7.5. Discussion 105
7.6. Conclusion 107
References 109

Part Two – **Extreme Situations** **113**

Chapter 8 The Project Front End: Financial Guidance Based on Risk **115**
Frédéric Gautier

8.1. The Project Front End: Learning Integration and Uncertainty 116
8.1.1. An Organizational View of the Project Front End 116
8.1.2. Integration of Knowledge in a Context of Uncertainty 118
8.2. Control Systems Based on Risk 120
8.2.1. Interactive Control Systems Directed to the Decrease in Uncertainty 120
8.2.2. The Multiple Contributions from Management Control Systems 122
8.3. Conclusion 124
References 125

Chapter 9 Lessons Learned from Sports Climbing: Some Disrespectful Discourse on Project Planning **127**
Valérie Lehmann

9.1. At the Foot of a Cliff 127
9.2. An Irrational Move 128
9.3. The Team before Anything Else? 129
9.4. Crossing the Golden Triangle 131
References 132

Chapter 10 Managing Extreme Situations in Fire and Rescue Organizations: The Complexity in Implementing Feedback **135**
Anaïs Gautier

10.1. Origins of Our Research Project and Focus on Analysis of Management Situations in Fire and Rescue Services 136
10.1.1. Sensemaking Applied to Management Situations 137
10.1.2. An Organization Stigmatized by Its Own History 137
10.2. The Organizational Framework of Feedback 138
10.2.1. The Regulatory Factor 139
10.2.2. The Structural Factor 139

10.2.3. The Cultural Factor 140
10.2.4. The Cognitive Factor 140
10.3. Conclusion 141
References 142

Chapter 11 Coordination Practices in Extreme Situations: Lessons from the Military 145
Cécile Godé

11.1. Coordinating in Extreme Situation: The Case of Air-to-Ground Operations in Afghanistan 146
11.1.1. Case Setting and Methodology 146
11.2. Coordinating in Extreme Situation through Bundles of Practices 152
11.3. Conclusion 154
References 155

Chapter 12 Developing Collective Competence in Extreme Project Teams: The French Special Forces Case 157
Tessa Melkonian & Thierry Picq

12.1. Conceptual Framework: Developing Collective Competence in Teams 158
12.1.1. The Concept of Collective Competence 158
12.1.2. Collective Competence in Project Mode 158
12.1.3. Studying the Development of Collective Competence: Issues and Questions 159
12.2. Project Management in Extreme Situations: The Case of the Special Forces 160
12.2.1. A Brief History 160
12.2.2. Specificities of the SF Commando Projects 160
12.2.3. Methodology and Specific Conditions of Study 161
12.3. Development and Mobilization of Collective Competence in the Special Forces 162
12.3.1. Strong Individual Expertise 162
12.3.2. The Combination of Different but Complementary Expertise 162
12.3.3. The Construction of a Shared Representation Based on Common Reference Systems and Language 163

12.3.4. The Capacity of Collective Improvisation 163
12.3.5. A Collective Memory 164
12.3.6. Subjective and Shared Commitment 164
12.4. Discussion: From Project Management to Management by Project 165
12.4.1. Contributions to Collective Competence in Projects 165
12.4.2. Contributions to the Dynamic of the Development of Collective Competence 166
12.5. Conclusion 167
References 167

Chapter 13 Situated Teams: Dropping Tools on Mount Everest 171
Markus Hällgren

13.1. Background 172
13.1.1. Temporary Organizations 172
13.1.2. Team Formation 173
13.1.3. Tool Dropping 175
13.2. Climbing Expeditions as Temporary Organizations 176
13.3. Methodology 177
13.4. The Mount Everest 1996 Disaster Revisited 179
13.5. Discussion 185
13.5.1. The Task-Situated Team 186
13.5.2. The Survival-Oriented Situated Team 188
13.5.3. The Situated Team Oriented toward Rescue 189
13.5.4. Toward an Increased Understanding of Situated Teams 190
13.6. Implications 191
13.7. Conclusion 192
Acknowledgments 193
References 193

***Part Three* – Lessons to Be Learned 197**

Chapter 14 Planning Risk and Cool Heads: Survival Conditions Required for Managing Projects 199
Jean Martel

14.1. Expeditions and Projects 200

14.2. Analogy between Expeditions and Construction Project Management 200
14.2.1. Initiating 201
14.2.2. Planning 201
14.2.3. Execution 202
14.2.4. Monitoring and Control 203
14.2.5. Closing or Ending 204
14.3. The Essential Qualities of a Project Manager 204
Reference 204

Chapter 15 Flexibility and Rigidity in Planning a Program: The Case of the Montreal Metro Renovation Project 205
David Brazeau

15.1. The Reno-System Program 205
15.2. Rigidity 206
15.3. Flexibility 207
15.4. Finding and Maintaining Balance 208
15.5. Conclusion 208

Chapter 16 Project Manager: Specialist or Generalist? 211
Benoit Lalonde & Maude Brunet

16.1. Description 212
16.1.1. The Main Constraints in Organizations 213
16.1.2. Specialist or Generalist 214
16.1.3. Organizational Enablers 214
16.2. Conclusion 215

Chapter 17 Project Management and the Unknown 217
Jean-Pierre Polonovski

17.1. Shared Characteristics of Expedition Projects and R&D Projects 217
17.2. Ambiguities of Project Management 218
17.3. An Indispensable Factor in the Success of a Project 219
17.4. Flexibility and Control 220
17.5. Management and Creativity 221
17.6. Conclusion 222
Acknowledgments 223

Chapter 18 Control and Flexibility: Which Balance Do We Mean? **225**
Danielle Desbiens

18.1. Which Balance Do We Mean? 225
18.2. The Project Environment 227
18.3. The Team Environnent 227
18.3.1. Managing a Team 228
18.3.2. Managing as a Team 228
18.3.3. Managing with a Team 229
18.4. An Individual's Environment 229
18.4.1. Competency 230
18.5. Conclusion 231
References 232

Conclusion **233**
Gilles Garel

Epilog **237**
Pascal Lièvre

Afterword: Looking for the Ordinary in the Extraordinary! **245**
Linda Rouleau

Index **251**

Foreword

Ann Langley

When Monique Aubry asked me to write a Foreword for this volume, I could not help wondering: Why me? I am neither a specialist in project management, nor am I addicted to extreme situations. My research sites are large and complex organizations rather than temporary organizations, and while I enjoy spectacular scenery, I like to remain close to civilization and view it from a safe distance. However, on reflection, perhaps I am the perfect reader for this volume. The subtitle of this book includes the word "lessons," suggesting that I might have something to learn from these experiences that are, for me, particularly strange. So what might I learn? Indeed, what have I learned from reading this book? In the next few paragraphs, I will try to answer these questions, hoping to stimulate the interest of other readers, whether they be specialists in project management, adventurers in search of the extraordinary, or, like me, ordinary professors of management seduced by the premise of this book that extreme situations can offer lessons for organizational life.

Indeed, the volume offers a rich variety of lessons and insights. On one level, what I particularly appreciate in this book is to be transported to unknown places and to experience vicariously some intense moments along with the participants (while safely ensconced in my armchair). The stories presented in this book are often richly described and deeply touching: for example, the episode with the dogs in Chapter 2 by Monique Aubry and Pascal Lièvre, Chapter 13 by Markus Hällgren about the Mount Everest climbers, and the adventure with the Neeposh clan in Chapter 6 by Nathalie Guérard and Anne-Marie Cabana. The reader feels the deep involvement of the authors in these experiences, and these stories deliver far more than any abstract academic interpretation could communicate. These authors are first and foremost accomplished storytellers.

However, stories without plots are never totally satisfying. To derive lessons from their accounts, the authors often tell us what they themselves learned from these experiences. For example, Aubry and Lièvre use the incident of the dogs and another incident on a drifting boat in the Antarctic to discuss the notion of ambidexterity in project management and show how exploration and exploitation before and after a project can interact, creating consequences that may be surprising and potentially valuable for project leaders. Similarly, Hällgren draws on the Mount Everest experience to introduce concepts associated with the spontaneous emergence of novel team relations in projects under stress. Underlying several of these contributions are the eternal tensions between planning and improvisation, between the predictable and the unknown, and between organized foresight and situated creativity.

The authors thus draw skillfully on extreme situations to suggest concrete lessons for project management in more mundane business contexts. They also find diverse ways to communicate their lessons. Some, such as Tessa Melkonian and Thierry Picq, propose to transfer knowledge directly, as in Chapter 12, where they explain how the French Special Forces develop collective competencies. Others use more traditional academic forms that involve presenting an empirical case study and analyzing it according to a theoretical framework from the literature on organizational theory or sociology to derive understanding, as in the case of the various comparative studies of Lièvre and colleagues (see Chapters 1, 3, 4, and 5). In Chapter 9, Valérie Lehmann innovates by offering us a kind of allegory on project management as rock-climbing. Others, such as Alain Grenier (Chapter 7) and Guérard and Cabana (Chapter 6) plunge the reader into the context with the use of photography.

The book also offers lessons on methodology that are particularly appealing to someone like me who is interested in qualitative methodology and the analysis of strategic practices in organizations. Chapter 5, by Géraldine Rix-Lièvre and Pascal Lièvre, proposes two complementary methods (i.e., observer participation and participant observation) that have been effective for their studies on extreme situations, and that might also be valuable in less unusual organizational contexts.

Dissatisfied with simply creating a series of disparate contributions with their own distinctive lessons, the editors of this book also had the brilliant idea of inviting a series of experts in project management—four practitioners and an academic—to present their own lessons from these studies of extreme experiences. This is a solid contribution to the book, placing in perspective the individual situations investigated in this book, integrating their lessons, and grounding them in organizational life, without denying their special character.

Finally, building on their own interests and experiences, all readers will find their own nuggets of insight in this book beyond those expressed explicitly. For my part, in reading this book, I was struck by how extreme situations reveal our vulnerability and interdependence with others. It seems also that isolation, the loss of habitual reference points, and situations of crisis accentuate the fluidity of interpersonal relations and contribute to the emergence of informal or collective leadership, a phenomenon that particularly interests me. It would be useful to examine whether the same types of setbacks and transformations in relationships show up in business organizations as they go through crisis. Finally, this book suggests to me that the study of project management would benefit from more detailed ethnographic work than is usually seen in business studies. In the book's Introduction, Christophe Bredillet also comments on the need to broaden research perspectives on projects and their management, which corroborates this thought.

In conclusion, I confess that this original and stimulating book also made me think about some of the large projects in which I have recently been involved. While I have never been in any physical danger, the feeling of uncertainty and apprehension that accompanies such large undertakings resonates nonetheless. The book thus touches a sensitive nerve, even in someone who is neither a specialist of project management nor of polar expeditions. I hope you enjoy reading this book as much as I did, and that you also find much to learn.

Preface

In June 2009 the University of Quebec at Montreal (ESG-UQAM) hosted a two-day research colloquium on the theme, "Lessons Learned from Project Management of Polar Expeditions." Researchers from France, Sweden, and Quebec met in Montreal to examine project management in the context of polar expeditions. This colloquium was made possible thanks to the Project Management Chair at ESG-UQAM and the Clermont Research Center for Project Management (University of Clermont and ESC Group, Clermont, France). The Clermont Research Center has been conducting research in the area of management under extreme conditions, which has included for 10 years research on polar expeditions (Lièvre, 2001). In June 2010, the *Project Management Journal* published a selection of eight articles presented at the colloquium in a special issue, "Project Management in Extreme Environments" (vol. 41, no. 3). This work collects the full and rich range of presentations given at the colloquium.

But why would researchers in project management hold such an event? The growing complexity of projects, as well as the risk inherent in certain types of projects, often renders inefficient traditional practices and processes that are based on the hypothesis that everything about a project is known at its inception. Project goals and the means to attain them are often known only gradually as a project unfolds. Under these circumstances, traditional practices and processes may not be sufficient. A new look at planning is necessary to give projects the flexibility they need for their entire length.

The innovation-based and knowledge-based economy that began in the 1990s (Nonaka, Takeuchi, & Umemoto, 1996; Cohendet, 2005; Foray, 2009) highlights enterprises' ability to constantly be managing innovative projects in what we call extreme conditions—conditions that are knowledge-intensive, constantly evolving, high-risk, and filled with unknowns (Lièvre, 2005; Lièvre & Gautier, 2009; Lievre 2007). Thus we put forth the hypothesis that projects in extreme environments can be a source for informing more traditional enterprise projects in the context of today's economy. In fact, for many years organizations have been forced to invent on the fly new management rules to conceive and carry out such extreme projects as polar expeditions, summiting high-altitude mountains, military and special operations, and combatting forest fires. A classic case studied in project management is the 1897 expedition by the Norwegian Fridtjof Nansen, who, after 10 years of preparation, set off to conquer the North Pole by letting his ship, the *Fram,* be carried by ocean currents. It is an example of a failed project whose goal was not reached, but it nevertheless made remarkable advances to its field, which was Arctic exploration, as well as to the field of salvaging an exploration. This expedition became a model of organization known as the Norwegian school of polar exploration, and it enabled the conquest of the Northwest Passage (1905) and the South Pole (1911). It is also possible to investigate and learn from contemporary polar expeditions, which we do in this book.

Investigating projects in extreme environments makes project phenomena easier for researchers to study because the context is rugged, team members' reasoning is pushed to the limit, and the risks inherent in these types of projects make the players quickly adjust their actions.

We believe that polar or mountain expeditions, or more generally, ones in any extreme situation, are ideal for research in project management. Elsewhere these projects are considered a class apart from the rest, as is the case in Scandinavian literature on project management. However, extreme situations act as a metaphor in the sense that they form a concrete image that helps to capture interactions that in other organizations seem more abstract. Extreme situations in management, such as polar expeditions, offer great potential for learning about management of unexpected and unpredictable situations.

For several years now, a movement toward renewal has been forming in the project management research community. Several arguments and proposals have been put forth to advance project management in the field of organizational theory as a legitimate research object *per se.* A common trend of these arguments is that there is not a single theory of project management. We acknowledge instead the existence of multiple facets based on a great variety of viewpoints. This variety can be seen as a characteristic of a field that is evolving and has not yet stabilized. In this dynamic context, which is being investigated, the search is for a global overview in which a variety of approaches can both be integrated and promoted to formalize a coherent model of project management.

The colloquium held in Montreal was a part of the movement of renewal by questioning certain traditional fundamentals concerning issues of philosophy, application, society, and methodology. The presentations at the colloquium covered a range of specific perspectives. Each presentation offered its own perspective, but together they formed a coherent and integrated approach to research.

Before getting into the book's content in-depth, we look at the Introduction by Cristophe Bredillet on the philosophical basics of project management. This piece reviews the ontological and epistemological fundamentals of project management. Bredillet uses this occasion to gather his pieces published when he was the director of the *Project Management Journal*. Based on Le Moigne ("Modeling to Understand—That Is, to Do Ingeniously," 2003), he proposes an approach to creating a meta-model. This model allows the paradoxes inherent in the dynamics of project management to be reconciled.

In Chapter 5, Rix-Lièvre and Lièvre define a new methodological approach to capture activities as they happen. The challenge is looking at an organization in action. In this way, a project comes into being as activities take place. The project observatory described in this chapter consists of a multimedia logbook and a device to observe practices. The researcher plays an active role. The device is a mini-camera installed on the researcher's eyeglasses. Image, sound, and all of the action are captured for analysis, along with the usual analysis of written documentation. Analysis involves the project participants reflecting detachedly on the different records constructed during the expedition. A narrative is constructed after the fact by the collective of participants and researchers.

The subject of Chapter 6, by Nathalie Guérard and Anne-Marie Cabana, is an expedition that does not take place in a polar region, but rather in Northern Quebec. It is the story of a Crie family's last traditional fishing trip in an area that will be submerged by the diversion of one Northern Quebec's great rivers, the Rupert. This exceptional project took place as part of an environmental impact study. Nathalie Guérard accompanied the Neeposh family daily during the entire course of the expedition. Her role as an observer and participant allowed her to capture the richness of the Neeposh's life. This moving story tells about their attachment to the land, their profound knowledge of it, and, most of all, how this all was part of the social and spiritual life of this community. The Neeposh family's goal in this project was to make known these deep values, which the environmental study coldly called an impact analysis.

To conclude the book's first part, Alain Grenier in Chapter 7 broaches the delicate subject of polar tourism and its impact on the environment, where there is a tension between being curious about an object and destroying it. He first places today's growing demand for eco-tourism in the overall debate about the environment. He questions how the words *adventure* and *expedition* are used in marketing by the tourism industry, and the roles these words play in our modern society. Grenier juxtaposes in his chapter two overall themes in nature tourism: valuing harmony with the surroundings and encouraging humans to conquer and dominate the environment. This chapter has insights invaluable to researchers fascinated by Polar regions.

The second part of this book covers extreme situations other than polar expeditions. Polar expeditions are a remarkable area of research, as we have already stated. However, all extreme situations are fertile grounds for better observing and understanding the hidden workings of project management.

Chapter 8, by Frédéric Gautier, starts this section on extreme situations with a stage that comes before the project: the pre-project phase in an uncertain and risky environment. Using a literature review, Gautier proposes to go beyond the current limits imposed by project management in developing new products or services into what he calls "a project's prehistory." The pre-project phase is at the same time both a phase for exploration and a phase for preparation. It is based on lessons learned and aims to state the problem at hand and to allow the project to get underway. The author proposes in this chapter new systems to pilot projects that favor analysis and risk management.

Chapter 9, by Valérie Lehmann, is on a different note entirely. It humorously discusses mountain and rock climbing. "Lecture de voie" is a technique currently popular in this sport, and its translation from French means "Reading the way." It is used to plan an ascent. But does this technique matter during the actual ascent? The chapter attempts to answer this question with reference to project management and rock climbing. And it does so with a good dash of humor.

Chapter 10, by Anaïs Gautier, is on reviewing experiences of rescue groups in extreme situations. Project management uses project assessments or post-mortems. In general, these processes allow the knowledge of individuals to be transformed into collective knowledge that can be distributed and leveraged. Gautier suggests a multidisciplinary and clinical approach to review experiences. This approach leads to a new dimension for learning that encompasses the organization and its context. This approach comprises four components: structural, cognitive, cultural, and regulatory. The author herself is a volunteer fire fighter, and she illustrates how this approach is used in fire and rescue services in France.

Chapter 11, by Cécile Godé, explores the mechanisms of coordination in extreme situations. In this case, the extreme situation is the French Air Force on its mission in Afghanistan. Innovative mechanisms were devised to call attention to individual and group competencies.

In Chapter 12, Tessa Melkonian and Thierry Picq propose an analysis of the components and dynamic processes that develop competencies on many levels (i.e., individual, group, and organizational).

With its provocative title, Chapter 13, by Markus Hällgren, is a profound reflection on what project management really is. Too often occupied with tools and processes, project teams often forget to focus their effort on the essential item, the team. Hällgren is Swedish and is a member of the Scandinavian School, which was the first to define a project as a temporary organization. His chapter follows in this line and emphasizes research on practices. In developing his approach, Hällgren examines a disastrous situation as recounted by its survivors, the 1996 Mount Everest expedition, during which a part of the team perished.

Not all projects place the lives of its participants in danger, fortunately, but here is a good example of an extreme situation that illustrates team phenomena. Markus draws lessons from it for researchers, practitioners, and teachers.

Finally, the book's third section brings together the different points of view presented at the round table that concluded the colloquium. The round table had five participants: four project management professionals and a psychologist who is also a professor of project management. This section contrasts with the preceding ones because it addresses questions about flexibility and control in relation to concrete situations in organizations. This perspective allows the problems taken up at the colloquium to be placed in the context of professional organizations. We believe this perspective is an indispensable component in understanding a complex phenomenon.

Gilles Garel and Pascal Lièvre conclude with a piece on the significance of the colloquium and this book. Linda Rouleau completes the book with lessons drawn from her own experience of the 2009 "Darwin" expedition to Patagonia.

This book, as well as the colloquium, attempts to be a dynamic exchange between researchers and professionals who deal with the management of projects and expeditions. We believe that combining these different perspectives can facilitate a new approach to manage projects better.

Polar expeditions and extreme situations are more than metaphors for managing projects. They are real experiences for understanding and transferring knowledge to other types of organizations. As the organizers of the colloquium and editors of this book, we are proud to share as much as possible these promising results from the work contributed to the colloquium. Our object is to encourage other researches to contribute through innovative approaches to the better understanding of project management dynamics. Following Bredillet's call to model in order to understand, this work is an attempt to recreate the individual, as well as the reflexive, actor in a paradoxical social context.

We have to thank all the authors who contributed to this book and participated in renewing project management by offering new and unexplored perspectives. We hope that the innovative approaches presented here can lead to a wider renewal in the field of project management.

– Pascal Lièvre
– Monique Aubry

References

Cohendet, P., Créplet, F., & Dupouët, O. (2006). *Les gestion des connaissances: Firmes et communautés de savoir.* Paris: Economica.

Foray, D. (Ed.) (2009). *The new economics of technology policy.* Camberley, UK: Edward Elgar Publishing.

Le Moigne, J. L. (2003). *Le constructivisme, Tome 3: Modéliser pour comprendre.* Paris: L'Harmattan.

Lièvre, P. (2001). *Logistique en milieux extrêmes.* Paris: Hermès.

Lièvre, P. (2005). *Vers une logistique des situations extremes, de la logistique de project du point de vue d'une épistémologie de l'activité d'une expédition polaire.* HDR University of Aix Marseille II.

Lièvre, P. (2007). *La logistique.* Paris: La Découverte.

Lièvre, P., et Gautier, A. (2009). Logistique des situations extrêmes: des expéditions polaires au service secours incendie. *Management et Avenir, 24,* 196–216.

Nonaka, L., Takeuchi, H., & Umemoto, K. (1996). A theory of organizational knowledge creation. *International Journal of Technology Management, 11*(7–8), 833–845.

Acknowledgments

We wish to warmly thank all authors who accepted our invitation to contribute to this book. It was, itself, a journey.

We also want to thank the Centre de recherche clermontois en gestion et management (Université de Clermont Auvergne and Groupe ESC Clermont France) and the Chair in Project Management at ESG-UQAM for their financial support in the organization of the colloquium. This colloquium was a unique occasion to bring together scholars from different backgrounds and countries around a common goal of making *extreme situations* a fruitful academic ground.

– Monique Aubry
– Pascal Lièvre

About the Authors

Monique Aubry

Monique Aubry, Ph.D., is a professor at the School of Business and Management, Université du Québec à Montréal (UQAM). She teaches in executive MBA and graduate project management programs. Her research interests focus on the planning process in extreme situations and on the organizing of projects and organizational design, more specifically on Project Management Offices (PMO). The results of her work have been published in major academic journals in project management and have been presented at several research and professional conferences. She is a member of the Project Management Research Chair (www.pmchair.uqam.ca) and the UQAM's Health and Society Institute. She is a senior editor for the *Project Management Journal*®. She is involved in the local project management community that oversees practices regarding organizational project management, where she promotes engaged scholarship and dialogue between professionals and researchers.

Email: aubry.monique@uqam.ca

David Brazeau

For over 25 years, David Brazeau, MGP, PMP, worked for the Montreal transit authority [Société de transport de Montréal (STM)] in the project management field. He has been a planning and control officer, a project coordinator, and a project manager. He has participated in and carried out several major projects for the STM. Moreover, he played an active role in the implementation of project management, several project management offices, and portfolio management within the organization. David holds a bachelor's degree in business administration from Université Laval and a master's degree in project management from Université du Québec à Montréal. He obtained his Project Management Professional (PMP) certification in 1994.

Email: dbrazeau55@gmail.com

Christophe N. Bredillet

Professor Christophe N. Bredillet is a professor of project management at Université du Québec à Trois-Rivières (UQTR) and is the Director of the joint committee/masters' programs network at the University of Quebec. He is the Scientific Director of the Société française pour l[1]avancement du Management de Projet (SMAP) and an adjunct professor at the Queensland University of Technology (QUT) Project Management Academy. He specializes in the fields of Portfolio, Program & Project Management (P3M). From 2012 to 2015, he was the Director of the QUT Project Management Academy. Before joining QUT, he was a Senior Consultant at the World Bank, and, from 1992 to 2010, he was the Dean of Postgraduate Programs and Professor of Strategic Management and P3M at ESC Lille. His main interests and research activities are in the field of Philosophy of Science and Practice in P3M, including the dynamic of evolution of the field; bodies of knowledge, standards, and their link with capability development; and capacity building, governance, and performance. He was the Executive Editor of the *Project Management Journal*® between 2004 and 2012. In 2012, he received the prestigious Manfred Saynish Foundation for Project Management (MSPM) – Project Management Innovation Award for his contribution to a philosophy of science with respect to complex project management.

Email: Christophe.Bredillet@uqtr.ca

Maude Brunet

Maude Brunet is a doctoral student in Management—with a specialization in Project Management—at the School of Management, Université du Québec à Montréal (ESG UQAM), under the supervision of Monique Aubry. Her research interests focus on the governance of major public infrastructure projects. More specifically, she is studying the Quebec (Canada) governance framework for major public infrastructure projects, how it has developed over time, and how it is applied to managing projects into practice. She has 10 years of experience in project management. She is currently a lecturer at ESG UQAM for the Master's Program in Project Management. She is also actively involved with the organization GP-Quebec, the association for public projects in Québec, and PMI-Montreal.

Email: maude.aumont.brunet@gmail.com

Anne-Marie Cabana

Anne-Marie Cabana, a graduate in biology, literature, and philosophy, has a very diverse background. She has worked in areas such as coastal fisheries and marine biology, physical oceanography, as well as the development and conservation of natural environments. Her mandates led her, among others, to Basse-Côte-Nord, where she shared the life of a fishermen community, and to Cayenne, French Guiana. Anne-Marie is currently employed by the Fisheries and Oceans Canada's Science Branch at the Maurice Lamontagne Institute, located in the Bas-Saint-Laurent.

Email: anne-marie.cabana@dfo-mpo.gc.ca

Danielle Desbiens

Danielle Desbiens, Doctor of Psychology (Ph.D.) and certified coach (PCC), has focused her teaching and professional practice on interpersonal behavior in the business environment, particularly in project management and supervisory roles. Prior to July 2009, Danielle was a tenured professor in Organizational Behavior and Human Resources Management at the School of Management at UQAM (ÉSG-UQAM). She also held several management positions at this University, and she has served on the boards of public and quasi-public organizations. She has published books and articles on the case method and project management teams. Her lectures cover the topics of leadership and supervision, self-management as a success factor, and coaching as a method of development and learning. As a consultant to senior and middle management, Danielle has developed expertise in coaching the next generation of business leaders.

Email: desbiens.danielle@uqam.ca

Gilles Garel

Gilles Garel is a Full Chair Professor of Innovation Management at the Conservatoire national des arts et métiers (Cnam) in Paris, France. He is also Professor at the Ecole polytechnique in Palaiseau, France. At the Laboratoire Interdisciplinaire de Recherche en Sciences de l'Action (LIRSA) at Cnam Paris, Gilles conducts research in the field of innovation and design management, in close collaboration with innovative firms and organizations. Former Professor at the School of Management at Ottawa University, Canada (now, the Telfer School of Management), Gilles Garel has published in a range of journals and has contributed papers and presentations at numerous conferences and seminars. He's notably a co-author of the recently published book, *The*

Innovation Factory (CRC Press/Taylor and Francis Group, 2016), with Elmar Mock, the Swatch co-inventor.

Email: gilles.garel@lecnam.net

Anaïs Gautier

Anaïs Gautier has a Ph.D. in Management Sciences (Aix-Marseille University). Currently, she directs the laboratory management and control organizations at the Centre for Studies and Interdisciplinary Research of Civil Protection at The National School of Firemen Officers (ENSOSP – Ecole Nationale Supérieure des Officiers de Sapeurs-pompiers). She is also an associate researcher at the Clermont Research Center for Management (CRCGM, EA 3849) in a program directed by Pascal Lièvre on managing extreme situations (Theme Human Potential, Organization, Innovation). In 2010, she wrote a thesis on the modalities of implementation of practical experience feedback within the fire and rescue services in France. She developed her research on the command and control of rescue operations during forest fires in the south of France. In particular, she created a volunteer firefighter expert officer function for the practice of feedback experiences during rescue operations. The purpose of this research is to develop a learning organization. She has operated on the highest level of training command and control (for forest fire operations) at the School of Application of Civil Protection (ECASC) since 2009. In addition, she has contributed to the development of the practice of feedback in the training of fire brigade officers at the ENSOSP since 2015.

Email: anaisgautier@hotmail.com

Frédéric Gautier

Frederic Gautier, a graduate of Sciences-Po Paris, is a professor of management at the IAE Paris Sorbonne Graduate Business School, a researcher at GREGOR (Groupe de Recherche en Gestion des Organisations), in charge of the center Rosks and Organizations, and a founding member of PRIMAL (Paris Research in Management and Law). Previously, he was a financial auditor in an international firm and a financial controller in the car industry. He defended his dissertation in 2002 at the University of Paris Panthéon-Sorbonne. This thesis was awarded by FNEGE. His research interests include the organization and management of projects, working closely with major European companies.

Email: gautier_frederic@orange.fr

Cécile Godé

Cécile Godé is currently full professor at Lyon 2 University, COACTIS Research Department, and associate researcher at the Research Center of the French Air Force (CReA). She is also at the head of the research chair "risk management and performance in SMEs and ISEs." Her privileged research fields are management and information systems management. Her current contributions focus both on decision/coordination practices and processes as well as experiential learning processes under extreme contexts. Her research has been published in academic journals such as *Système d'Information et Management, The International Journal of Technology and Human Interaction, Management International, European Management Journal,* and the *Scandinavian Journal of Management.* She also authored the book, *Team Coordination in Extreme Environment: Work Practices and Technological Uses under Uncertainty* (John Wiley & Sons, Inc./ISTE Ltd., 2016).

Email: cecilg@orange.fr

Alain A. Grenier

Alain A. Grenier is a professor of nature-based tourism and sustainable development in the Department of Urban and Tourism Studies at the School of Management/Université du Québec à Montréal (ESG UQAM). A doctoral graduate in sociology from the University of Lapland on the Arctic Circle in Finland, Alain's research focuses on the contradictive relationships between people and the natural environment in the context of tourism planning and sustainable management in nature and polar regions. Dr. Grenier's studies on adventure tourism have leaded him both to Antarctica (1993, 1994–1995) as well as across the Northeast Passage (1998) and the Northwest Passage (1999, 2010, 2013). A former editor of the French language academic journal of tourism, *Téoros* (2008–2014), Professor Grenier is currently the head of the master's degree program in tourism at the Université du Québec à Montréal.

Email: Grenier.alain-adrien@uqam.ca

Nathalie Guérard

Nathalie Guérard, senior technician in wildlife management, has many years of field experience, mainly in Quebec (in the northern regions) and abroad. Specializing in aquatic fauna, she has developed a recognized expertise in the characterization of fish habitat and sampling strategies. She has been closely involved in numerous impact assessments and environmental monitoring programs related to major hydroelectric projects in Canadian watersheds. In 2006, as part of the social impact assessment related to the diversion of the Rupert River, she told the journey of a Cree family across their ancestral lands. Nathalie has established privileged and lasting relationships with the Neeposh, who she accompanied for several weeks along a fishing course. She shared their daily life and took part in their traditional hunting, fishing, and the establishment of camps.

Email: nathalie.guerard@wspgroup.com

Markus Hällgren

Markus Hällgren is a professor of management and head of research at the Umeå School of Business and Economics, Department of Business Administration, at Umeå University, Sweden. Hällgren is the founder and organizer of the international, interdisciplinary research program, "Extreme Environments – Everyday Decisions" (www.tripleED.com), as well as co-founder and organizer of the international network, "Organising Extreme Contexts" (www.organizing extremecontexts.org), together with Professor Linda Rouleau, at HEC Montreal. Hällgren's research interests lie within organization theory, temporary organizing, and extreme contexts. Within that broad topic, he is particularly interested in sensemaking, organizational routines, team dynamics, and the methodological and theoretical challenges and opportunities of a practice-based approach.

Email: markus.hallgren@umu.se

Benoit Lalonde

Benoit Lalonde is president and founder of GPBL Inc. He has developed a methodology and approach to project management that is based on the PMI's best practices as well as his 20 years of professional experience. He is devoted to offering project management consulting and training at organizations of all sizes. As an enthusiastic and dynamic participant in his profession, he knows how to effectively transmit his passion for project management to those he works with or teaches. Constantly on the lookout for new developments in his area of expertise, he innovates by adapting his range of project management products and services to the latest trends in the sector. As a specialist in project management, he is an accredited member of the Project Management Institute and a certified PMP (Project Management Professional), a CPM (Certified Project Manager), an OPM3 Assessor and Consultant, and a PMI-RMP (Risk Management Professional). He is the co-author of *Computerized Project Management: Microsoft Project 2007*, and he applies an Integrated Project Management Methodology (IPMM)—a simple and multimodal approach—to his client interventions. GPBL is a company that distinguishes itself in three distinct market segments: consulting, training, and project management resource placement. Conscious of today's modern and innovative project environment, GPBL has developed a gamut of products and services that are simple and easily adaptable to the needs and context of its clients, no matter their industry. An expert in management by project, GPBL contributes to the improvement of organizational performance in management by offering quality services to the numerous organizations it works with.

Email: lalonde.benoit@gpbl.ca

Ann Langley

Ann Langley is a professor of strategic management at HEC Montréal, Canada, and holder of the Chair in Strategic Management in Pluralistic Settings. Her research focuses on strategic change, interprofessional collaboration, and the practice of strategy in complex organizations. She is co-editor of the journal, *Strategic Organization,* and co-editor, with Haridimos Tsoukas, of a book series, *Perspectives on Process Organization Studies,* published by the Oxford University Press. She is also adjunct professor at Université de Montréal and the University of Gothenburg.

Email: Ann.Langley@hec.ca

Marc Lecoutre

Marc Lecoutre, Ph.D. in Sociology, is a professor of organization theory and management at the Groupe ESC Clermont and a researcher at the CRCGM (EA 3849, Centre Clermontois de Recherche en Gestion et Management) and at the LISE-CNAM (UMR 3320, Laboratoire Interdisciplinaire de Sociologie Economique, Paris). His research focuses on the role of social networks in collective action and the management of organizations. In 2008, Dr. Lecoutre and Pascal Lièvre co-authored, *Management et réseaux sociaux. Ressource pour l'action ou outil de gestion [Management and Social Networks: Resource for Action or Management Tool?]* (Hermes & Lavoisier, 2008).

Email: marc.lecoutre@esc-clermont.fr

Valérie Lehmann

Valérie Lehmann is a professor of project management for the School of Management at the Université du Québec à Montréal (ESG UQAM) in Canada. She holds a Master's Degree in Communication, an MBA, and a Ph.D. in Administration. She was a project manager for 12 years. Her research, consulting, and publishing activities focus on stakeholder, change, and knowledge management and co-innovation systems as Living Labs. She belongs to several Chairs, as well as to the Chair in Project Management ESG UQAM. Currently, she participates in several interdisciplinary research programs on collective projects and social acceptability. She co-edited two collective books: *Communication and Large Scale Projects* (PUQ, 2013) and *Change and Large Scale projects* (PUQ, 2015). Presently, she is working on a new book, *Living Labs and Collective Approaches for Societal Projects*, to be published in 2017.

Email: lehmann.valerie@uqam.ca

Pascal Lièvre

Pascal Lièvre is a full professor in management science at Clermont University, EA 3849 CRCGM. He received a Ph.D. in Production Economics from the University of Lyon-II. Since 2000, he has been in charge of a research program on the management of extreme situations at the Centre de recherche clermontois en gestion et management (CRCGM). He has published 7 books and 40 academic articles on this topic.

Email: pascallievre@orange.fr

Jean Martel

Jean Martel, who completed his Diploma of College Studies in Architecture (1986) and his Bachelor of Engineering and Construction Management (1989), has accumulated over 25 years of expertise in managing complex projects and in property management. He is currently Director of Project Management Teams for the modernization of a university hospital in Montreal (New CHUM)—a project that has cost 3.5 billion CAD. Jean Martel also participated in several unconventional projects, including investment projects such as Six Flags in Montreal (100 million CAD), construction of the 1000 de la Gauchetière (250 million CAD), and several others, thereby providing him with experience in most areas of construction. He acquired expertise in this field through his work as electromechanical director of property management. In 1996, he was awarded the first energy-saving award of the Quebec Association of Energy Management (AQME) in the "large buildings" category. His current involvement in sustainable building (numerous LEED projects) and within the PMI (as the director of community of practices in construction) rounds out his practical experiences in the field of construction, incorporating ecological approaches and project management.

Email: jmartel@bgsemtec.com

Tessa Melkonian

Tessa Melkonian is an associate professor of organizational behavior and management at EMLYON Business School, France. She received her Ph.D. in Management from the University of Paris II. Her research interests include the role of justice and exemplarity in employees' reactions to change, determinants of the cooperation of employees in the context of Mergers and Acquisitions (M&A), and collective performance in extreme teams. She has published several articles in international journals (*Human Resource Management, Journal of Business Ethics, International Studies in Management and Organization, International Journal of Project Management, Project Management Journal*®) as well as professional articles and book chapters. In 2013, she received a French Academic Research Award for her research on the role of justice and exemplarity during M&A.

Email: melkonian@em-lyon.com

Thierry Picq

Dr. Thierry Picq is the Academic Dean at the EMLYON Business School (since 2014) and Professor of Human Resources Management (HRM) at EMLYON. Previously, he held the position of head of the HRM Department (four years) and was Associate Dean of pedagogical innovation (three years). He teaches both students (Master's Degree in Management and MBA) and executives in the fields of organizational behavior, team management, and project management. Thierry has five years of experience as a consultant, during which time he carried out many projects in organizational change and management development for large French and international companies. Thierry has a Ph.D from Grenoble University. He managed many research projects on project-based organization, high-performance project teams, managerial innovation, collaboration and interorganizational projects, change management, collective competencies and intelligence, and management in extreme contexts. He has published many books and articles in a large variety of French and international professional and academic journals, including the *International Journal of Project Management*).

Email: picq@em-lyon.com

Jean-Pierre Polonovski

Jean-Pierre Polonovski holds a Ph.D., MBA, and PMP certification and has always worked in the field of technology. He has managed several hardware and software projects and programs, including the European Supercomputer project. He has experience with the delicate problems involved in managing research and development projects, in which unknown factors often far outweigh known factors in the planning phase. He is also an entrepreneur and has created several technology companies, managing the development of software products, which have had a significant impact on the operation of the organizations that have introduced them.

Email: jppolonovski@CodeObject.com

Michel Récopé

Michel Récopé is a senior lecturer at Blaise Pascal University (Clermont-Ferrand, France) and a member of the *Activité, Connaissance, Transmission, Éducation* Laboratory. He works within cognitive anthropology, bodily practices, and the study of activity (practical knowledge, personal experience, and sensemaking). He is particularly interested in sensibility as an instance of integrating cognition, affectivity, and motricity.

Email: Michel.Recope@univ-bpclermont.fr

Géraldine Rix-Lièvre

Géraldine Rix-Lièvre holds a Ph.D. in Sport Sciences. She is full professor at Blaise Pascal University (Clermont-Ferrand, France) and a member of the laboratory ACTé (*Activité, Connaissance, Transmission, Éducation Laboratory*), in which she coordinates a thematic focus on "asymmetric interactions dynamic." She studies the embodied dimensions of practices, especially those of polar expeditions and of refereeing situations and has patented new methods that make these dimensions of human experience explicit. She has contributed to several reviews: *Knowledge Management Research and Practice,* 2008, no. 6; *Revue d'anthropologie des connaissances,* 4(2010/2); *@ctivités,* 2014, 11 (1); *Recherches qualitatives,* 2014, Vol 33 (1); *Sociologie du travail,* 2015, no. 57.

Linda Rouleau

Linda Rouleau is a professor in the Department of Management at HEC Montreal. Her research focuses on micro-strategy and strategizing in pluralistic contexts. She is also researching the strategic sensemaking role of middle managers and leaders. In the last few years, she has published in peer-reviewed journals such as the *Academy of Management Review*, *Organization Science*, *Accounting, Organization and Society*, *Journal of Management Studies*, *Human Relations*, and more. She is co-responsible for the GéPS (Study Group of strategy-as-practice, HEC Montreal). She is also leading an international and interdisciplinary network on "Organizing Extreme Contexts" (www.organizingextremecontexts.org).

Email: linda.rouleau@hec.ca

Introduction

Blowing Hot and Cold on Project Management: A Little Essay on Integrative Onto-epistemology

Christophe N. Bredillet

A and Ω (so far. . .)

More than 20% of global economic activity takes place as projects, and in some emerging economies it exceeds 30%. World Bank (2012) data* indicate that 22% of the world's $48 trillion gross domestic product (GDP) is gross capital formation, which is almost entirely project-based.† All included, project-based

* From World Bank Indicators website, http://data.worldbank.org/indicator/NE.GDI.TOTL.ZS, accessed March 31, 2012.

† "Gross capital formation (formerly gross domestic investment) consists of outlays on additions to the fixed assets of the economy plus net changes in the level of inventories. Fixed assets include land improvements (fences, ditches, drains, and so on); plant, machinery, and equipment purchases; and the construction of roads, railways, and the like, including schools, offices, hospitals, private residential dwellings, and commercial and industrial buildings. Inventories are stocks of goods held by firms to meet temporary or unexpected fluctuations in production or sales, and 'work in progress.' According to the 1993 SNA [System of National Accounts], net acquisitions of valuables are also considered capital formation." (http://data.worldbank.org/indicator/NE.GDI.TOTL.ZS, accessed March 31, 2012)

activities represent about 50% of the global economy. This situation is even stronger in potentially high-growth countries (e.g., BRICS & Africa according to the report World Economic Outlook (WEO)—Rebalancing Growth—April 2010, ©2010 International Monetary Fund). Considering these facts leads to making the assumption that competence and capacity building in project management at every level (individual, team, organizational, societal, . . .) is a key aspect for sustainable socioeconomic performance (Crawford, 2007; Gareis & Huemann, 2007; see also http://www.pmiteach.org/why_teach_project_management, accessed August 18, 2013). Educational programs in project management have grown rapidly during the last three decades to fill the need for competence (Atkinson, 2006; Umpleby & Anbari, 2004). It is therefore necessary for project management to be developed as a rigorous academic field of study. This is essential so that rapid economic development, so dependent on project management, can be supported by sound theory and not just case history of doubtful rigor. Modern project management started as an offshoot of operations research, with the adoption of optimization tools developed in that field, and some members of the community have continued to present it as such. However, in this introduction, I wish to demonstrate that project management has now grown into a mature academic discipline of some diversity and complexity, and should be apprehended with an integrative onto-epistemological perspective. Several schools of thought in project management can be identified, and project management is increasingly drawing on and making contributions to research in other fields of management (Anbari, Bredillet, & Turner, 2008; Kwak & Anbari, 2008; Söderlund, 2011). In this way, project management is becoming substantially different from operations management, which continues to emphasize the application of optimization models, tools, and techniques to the analysis of production processes (Slack, Chambers, & Johnston, 2006).

1. Project Management, A Recognizable Field of Study?

Audet (1986) defines a knowledge field as

> . . . the space occupied by the whole of the people who claim to produce knowledge in this field, and this space is also a system of relationships between these people competing to gain control over the definition of the conditions and the rules of production of knowledge (p. 42).

Who does claim to produce knowledge within the field?

The users. In the early days of modern project management in the 1950s, the development of knowledge was led by the users (Morris, 1997).

The rise of professional associations and agencies. In the 1980s, leadership of the development of knowledge (i.e., bodies of knowledge, competence frameworks, and standards) was taken over by such professional associations as the Project Management Institute (PMI), the United Kingdom's Association for Project Management (APM), and the International Project Management Association (IPMA). They needed to develop bodies of knowledge to support their certification programs. The focus of this work continued to be very practitioner-based and -oriented, and so it did not always adhere to recognized standards of academic rigor. Other actors were active as well in this development: industries, sectors, national and international agencies, and non-governmental organizations (NGOs), for example, including the IS/IT industry, the construction industry, the World Bank, the United Nations, and defense/aerospace sectors (e.g., the National Aeronautics and Space Administration, the European Space Agency, etc.).

And then came academia. It is only over the last 10 to 15 years that universities and other academic research institutions have begun to provide leadership. The first academic research conference in project management, the biennial International Research Network for Organizing by Projects (IRNOP) conference, was initiated in 1994. The PMI started holding its biennial research conference in 2000, and the annual European Academy of Management (EURAM) conference has had a project management track since its inception in 2001, the Academy of Management having started timidly, and in a way ironically, via the Operations Management division in 2008. I could add to this the formal and informal development of research networks, academic and practitioner conferences, workshops and seminars, and the way they are interrelated and interact through researchers, practitioners, and institutional relationships (professional bodies, various "professional" organizations, national and international research agencies, and academic organizations) in order to "create knowledge." Examples of these include the PMI Research Community, IRNOP, EURAM, the Academy of Management (AoM), and the European Institute for Advanced Studies in Management (EIASM), to mention a few. With the academic community playing a primary role, in conjunction with the professional bodies and in relations with industries, in the rigorous development of the discipline, we can state that project management, although a relatively young field of study, deserves to be acknowledged as an academic discipline. Exemplifying this, the inclusion of the two leading academic journals in the field, the *International Journal of Project Management* and the *Project Management Journal*, in the Social Science Citation Index (SSCI) is an important step forward. Thus, according to Audet's above definition, we can acknowledge the emergence of an identifiable knowledge field.

1.1. A Place of Evolution and Revolution

The field is recent and is evolving. Initially, in the late 1960s, advanced study in project management in universities was located in schools of engineering or construction, and then in schools of computing. So it was viewed as a technical subject. More recently, in the 1980s, project management has also been incorporated into schools of business or management, and so is now gaining recognition as a branch of management. The evolution of the body of knowledge is evidenced further by the numbers of papers and books (with regard to quantity at least), and the diversity of themes, e.g., citing techniques from the psycho-sociology of temporary groups through knowledge creation and organizational learning to strategic management. In addition, the field is currently characterized by this abundance of initiatives, updates, and development of standards—and related competence frameworks and credentials—at various levels (project, program, portfolio, maturity models, etc.) and in various areas (risks, contracts, scheduling, etc.). However, the field is also a place of revolution, inaugurated by a growing though still narrow subdivision within the project management community, where the existing positivist paradigm has ceased to function adequately in the exploration of nature (e.g., Cicmil, Williams, Thomas, & Hodgson, 2006; Cooke-Davies, Cicmil, Crawford, & Richardson, 2007; Winter, Smith, Morris, & Cicmil, 2006).

1.2. Has Anyone Found a Paradigm Out There?

At this stage, I could argue that the field is in a pre-paradigmatic phase according to Kuhn's sense (1970): There is no consensus on any particular theory, though the research being carried out can be considered scientific in nature. The current phase of development of the field is characterized by several incompatible and incomplete theories and perspectives (Bredillet, 2010; Söderlund, 2011). As a young discipline, the theoretical foundation of the field is still in its early stages of development. A number of authors have indicated that development of a theory of project management is important to progress in the field and possible connections with other disciplines or knowledge fields (Artto, Martinsuo, Gemünden, & Murtoaro, 2009; Bakker, 2010; Bredillet, Tywoniak, Hatcher, & Dwivedula, 2013a; Cicmil et al., 2006; Meredith, 2002; Morris, 2013; Sauer & Reich, 2007; Söderlund & Geraldi, 2012; Turner, 2006; Walker, Cicmil, Thomas, Anbari, & Bredillet, 2008). A mutual improvement and some kind of cross-pollination should result from these works. This supports the need for a plurality of perspectives, as we have not yet any "grand unified theory." A particular perspective, if valid in a specific area, cannot produce answers to every type of problem or in any type of situation. If the actors in the pre-paradigmatic community eventually gravitate to one of these conceptual frameworks and ultimately to a widespread consensus on the appropriate choice of methods, terminology, and what kind of experiment is likely to contribute to increased insights, then the phase of "normal science" begins.

At the same time, considering, for instance, the "9 Schools of Project Management" (Bredillet, 2007), the "Complexity" (Cooke-Davies et al., 2007), the "Rethinking PM" (Winter et al., 2006), and the "Reconstructing PM" (Morris, 2013) researches, I could argue that we are in a paradigmatic shift phase, moving from the classical dichotomy modernist vs. postmodernist perspectives (Boisot & McKelvey, 2010) to a new one, i.e., a more balanced one combining positivism, constructivism, and subjectivism, enabling us to address complexity, uncertainty, and ambiguity, because the classical dichotomous thinking is not working anymore. However, as rightly noticed by Boisot and McKelvey, it can be assumed that an integrative approach providing a bigger picture could help us to discover how patterns within a specific perspective may support the general theory quest through scale-free phenomena (Boisot & McKelvey, 2010, p. 437). As a consequence, assuming that the project management knowledge field exists and is in a pre-paradigmatic or paradigm-shift phase, it is not surprising. Furthermore, I argue that many applications of project management are done without questioning the deep nature of projects. What does a project manager actually do in a context of activity or situation named "project"? What is project management in a given context, according to a specific perspective (ontological consideration)? On which epistemological foundations can we build the project management field? Which hypotheses apply to the field? What are the consequences on the development and use of theories, concepts, methods, and techniques? So many delicate questions calling for a need of clarification of the ontological and epistemological foundations of the field!

2. Ontological and Epistemological Issues and Considerations

After Polanyi (1958), and in line with Boisot and McKelvey's view (2010), I propose an alternative ontological perspective both to Parmenidean "being" and Heraclitean "becoming" and an alternative epistemic position to positivism, constructivism, and subjectivism. I have no intention to separate personal judgment and deliberation, in reference to Aristotelian phronesis (Bredillet et al., 2013a), from scientific method.

To paraphrase the construction of the famous ontological argument, I would state that:

1. The concept of anticipation of an expected result of an action to be undertaken is inherent to human beings.
2. "Project(ing)" (etymologically, Latin "projectum" from "projicere," i.e., "throwing something forth") is thus consubstantial to humanity.
3. Therefore project management, the science and art to "throw forth," i.e., to anticipate the future and thus to cope with the radical uncertainty of the future, turning "our eyes away from the blinding light of eternal certitude towards the refracted world of turbid finitude" (Long, 2002, p. 44), generating what Bernstein has named "Cartesian anxiety" (Bernstein, 1983, p. 18), does exist.

I argue that, especially in project management, knowledge creation and transfer are linked to intelligent action, "ingenium" (Vico, 1708), and have to integrate both classical scientific aspects and "fuzzy" or symbolic (etymologically "throwing things together," quite interestingly) aspects.

2.1. The Ontological Characterization of the Fundamental Nature of Project Management!

A "reality" can be explained according to a specific point of view or perspective and also can be considered as the symbol of higher order and a more general reality (for example, a two-dimensional form can be seen as the projection on a plan of an *n*-dimensional figure) (Guénon, 1986). I argue that the "demiurgic" characteristic of project management involves seeing this field as an open space, without "having" (Have) but rather with a *raison d'être* (Be), because of the construction of Real by the projects. It could be considered to be a fundamental explanation of the pre-paradigmatic or paradigm-shift nature of this field [see Kuhn (1970) previously]. However the dominant paradigm, the source of well-established theory(ies), is *not* to find.* The deep nature of project management

* Dogan (2001) addresses the question, "Is scientific progress in the social sciences achieved mostly by steady accretion or mostly by abrupt jumps? For Thomas Kuhn, who devised the concept 'paradigm,' there are no paradigmatic upheavals in the social sciences. For him the use of this term in these sciences is not justified. Three arguments can be advanced against its polysemic use or abuse. In contrast with the universal truth in the natural sciences, contextual diversity and social change are two important parameters in all social sciences. In political science, sociology, and economics, progress is achieved by cumulative knowledge, by the adding of successive layers of sediments. The third argument is the pattern of mutual ignorance among great social scientists. In the social sciences, theoretical and methodological disagreements are beneficial to the advance of knowledge." (Dogan, 2001, p. 11023)

implies that this paradox of being built on moving paradigms reflects the diversity of the creation process by itself.

The nature of "project management" is thus composed of both:

- *Quantitative aspects* (Have—being ontology placing emphasis on permanent and unchanging reality), dependent on the positivist and constructivism paradigms, where reality is considered to exist independently of consciousness
- *Qualitative aspects* (Be—becoming ontology placing emphasis on change and emergence), dependent on the subjectivist paradigm, where meaning is imposed on the object by the subject

2.2. Epistemic Integration

Project management, the construction of the World or Reality by the "projects," as a knowledge and action field (knowing and acting are inseparable in the performance of project management, as discussed by Bredillet et al., 2013a), is both an art and a science, in their dialogic *and* integrative dimensions, and thus according to the three epistemological approaches:

- *The positivist epistemology* (materialist—quantitative—Have): "The relation of science to art may be summed up in a brief expression: From Science comes Prevision, from Prevision comes action" (Comte, 1855, p. 43).
- *The constructivist epistemology* (immaterialist—qualitative—Be-Have), with two hypotheses of reference as underlined by Le Moigne (1995).
- *The subjectivist epistemology* (immaterialist—qualitative—Be): If we follow Searle (1997), any value judgment is epistemically subjective.

2.3. Project Management as a Complex Integrative Field

Based on previously discussed considerations, I would metaphorically qualify the project management field as the place of a mirror (Bredillet, 2004) used for intelligent action and reflection *and* reflexivity, actualizing creation of values (for people, organizations, and society). This is in the realm of complexity (Richardson, 2005), ambiguity, and uncertainty of interactions between multiple factors and actors, each of them having a specific time horizon and occupying a specific place, performing a specific role and, where it is helpful, transposing one experience to other analogical contexts and situations (Gentner, 1983). This work, the study of project management as a complex integrative field, is supported by complexity science, systems science, as well as the "relational"/"connectionist" perspective in social sciences (Bredillet, Tywoniak, & Hatcher, 2013c). Interestingly, it reflects the outcomes of research studies that call for new perspectives for (project) management (e.g., Andriani & McKelvey, 2009; Cooke-Davies et al., 2007; Hodgson & Cicmil, 2006; Jackson, 2003; Leybourne, 2007; Maylor, 2006; Stacey, 2010, 2012; Williams, 2002).

"Management". . . Kurtz & Snowden (2003) question the three basic assumptions that pervade the practice and the theory of decision making and thus the translation of an organization's mission into practice: assumption of order, assumption of rational choice, and assumption of intentional capability.

. . . of project. Boutinet (1996) suggests an anthropological approach for the "project" demonstrating superbly the polysemous nature of the concept of project, and the tensions and paradoxes involved by this polysemous nature (Boutinet, 1997). He shows that the project and its management can be understood as means to realize some very diverse ends or finalities. Boutinet's approach is fully coherent with the integrative onto-epistemological perspective suggested here. Through projects, man builds reality, as highlighted by authors such as Declerck, Debourse, & Declerck (1997). The management of projects is a process of "naming" [the name given to an entity reflects its quality (Hacking, 2002), place of the mirror, . . .], of revelation, of creation. The management of a project by its mode of deployment within the ecosystem project/organization/context implies a systemic vision, an "intelligent action," "ingenium," "this mental faculty which makes possible to connect in a fast, suitable and happy way the separate things," as stated by Le Moigne (1995), quoting Vico (1708).

A need for complexity. . . I concur with Kurtz and Snowden (2003) and would argue that management of projects needs to be understood as a complex discipline because it aims to deal with complex reality. A number of studies and their results support this proposition, such as the law of requisite variety (Ashby, 1958; for an application to social sciences, see, e.g., Andriani & McKelvey, 2009; Boisot & McKelvey, 2010). This implies that it is important to plan and accept for many states for the knowledge (and action) field (situations, perspectives) and many misunderstandings (see the role of conventions that follow).

. . . and simplicity! Project management also needs to be simple, as far as its principles are concerned (again, see the role of conventions that follow): As white light is transformed into multiple colors through a prism, project management applications may be seen as coming from some general principles (Andriani & McKelvey, 2009). In France, it is worthwhile to highlight the work on "meta-rules" pioneered by Jolivet (2003) in relation with the "Club of Montréal"—this author has developed the "meta-rules" approach since the late 1970s within

Spie-Battignoles—that exemplifies very cleverly such perspective. Unfortunately, his work was probably too pioneer and requested too much "ingenium" to be considered by the global academic community and by professional bodies. . . .

These considerations on the different perspectives embodied in the concept of project management, on the polysemous nature of the concept, and consequentially on the underlying integrative perspective consubstantial to the concept of management of projects and its paradoxical and nontraditional logic, lead me to illustrate some implications for the understanding (theory) and practice using project management standards as an example.

3. "Modeling to Understand" That Is to Do Ingeniously!

This title is the translation of the title of an editorial of the "Réseau Intelligence de la Complexité" (Network Intelligence of Complexity, May 2004). Indeed, what to do—how to act, create, and transfer knowledge?—in front of the complexity of situations and contexts both addressed by and part of the project management field? Knowing and acting in complex situations involves "modeling to understand" that is to do ingeniously (Le Moigne, 2003). According to a complexity and systems thinking perspective, acting and learning are inseparable. As Morin (1985, p. 232) said so well, "Every knowledge gained on knowledge becomes a means to highlight the knowledge which enabled to gain it. We can then find a way back to the one-way 'epistemology—science'" (our translation).

3.1. Modeling Principles

This modeling approach is well grounded in sound theoretical organizational frameworks. With a project management perspective, we can say the approach is about designing a contextual structure that:

- Provides a privileged and "situated" place for individuals, project managers, and stakeholders to act and learn (Social Learning Cycle, Boisot, 1998; Houde, 2007)
- Facilitates this praxis through a specific meta-method, based on a logic "action and act" (Von Mises, 1976, chap. 1, §6) fully congruent with the notion of "ingenium" and an epistemo-praxeology (Von Mises, 1981; Menger, 1985)
- Enables to generate a specific convention (Gomez & Jones, 2000; Gomez, 2006), on the basis of the two formers points, as well as some kind of stability to cope with uncertainty and ambiguity in a given project's complex situation

The modeling approach helps to create a coherent or dissonant framework of symbols, promoting adequate dynamic management practices (e.g., standards), i.e., action, knowledge creation, and transfer, while being conscious of rational voids.

3.2. Standard as a Convention: From "One Best Way" to "Ingenium"!

For the Project Management Institute, a standard is "a document, established by consensus and approved by a recognized body, which provides, for common and repeated use, rules, guidelines or characteristics for activities or their results, aimed at the achievement of the optimum degree of order in a given context" (http://www.pmi.org/PMBOK-Guide-and-Standards/Standards-Overview.aspx, accessed August 18, 2013). This view of standardization is rooted in the classical framework proposed by the International Standard Organization (ISO) (Brunsson, Jacobson, and Associates, 2000) and as shown by Bredillet et al. (2013c), and is supported by the classical economics "resource-based view" perspective, and equilibrium- and process-based views in terms of organizations theory. Using Jackson's (2003, p. 18) problem context matrix, we see that this view is well suited for contexts where actors or agents have a unitary relationship, i.e., they "have similar values, beliefs and interests," and where the system to deal with can be either simple or complex.

However, when situations involved participants with pluralist relationships, i.e., "basic interests are compatible but they do not share the same values of beliefs" (Jackson, 2003, p. 19), the classical perspective is not adequate any more and requires a different perspective. I would like to provide an alternative possible view of the principles and characteristics underlying what could be a standard in order to be congruent with the previous development and based on the modeling principles introduced above. This view embodies the classical perspective as a particular case of a broader perspective. Conventions theory, challenging and complementing the assumptions of classical economics on which traditional standardization is rooted, offers us both a theoretical framework and a method for "designing" models, enabling us to understand the systemic dimension and dynamic structure of standards seen as a special case of conventions (Bredillet, 2003). Where the classical perspective focuses on consensus and order imposed by some kind of transcendental and pre-existing normative orientation, common norms and values, the conventionalist perspective focuses on the coordination of relationships and cooperation between agents with or potentially with divergent interests and values. Thus standards, i.e., models, as conventions, are not deliberately designed from "out there" and used normatively but are dynamically emerging from some self-producing systems of rules and interpretation shared by adopters (Bredillet et al., 2013c).

What constitutes a convention? Gomez & Jones (2000, p. 700) answer: "A convention is a social mechanism that associates a rational void, i.e., a set of non-justified norms, with a screen of symbols, i.e., an interrelation between objects, discourses, and behaviours." We easily see that the traditional definition of standards is a particular case of this broader definition.

From this, several important consequences can be drawn and discussed, showing the richness of this approach (conventionalist lens and modeling) with regard to acting and knowledge creation and transfer in the field of project management and in the context of project situations.

The morphology of a governing system and its dynamic of evolution. To understand how conventional systems of rules are constituted and how they modify and transform themselves, we need to consider two characteristics of a convention: morphology and complexity.

Morphology. Using the work of Boltanski & Chiapello (2005), Boltanski & Thévenot (2006), and Le Moigne (1990), Gomez proposes a general framework describing the morphology of any convention (Gomez, 2006, p. 224). We must note that a convention is not an *ex nihilo* construction, but is the result of individual agents' behaviors, accepting it because they are convinced that others are accepting it. A convention conveys the conviction about its own generalization. This can be done either by statement (discourse, narrative) or by material apparatus. Thus a convention is described by two sub-subsystems constituting a general referential enabling comparison between different conventions or governing systems:

Statement

1. The higher-order principle provides the convention purpose, considered as "good," "positive," in the conventional rules.
2. The distinction provides the typology of the different adopters of the convention, hierarchical relations between them and their relative place.
3. The sanction provides the motives for inclusion or exclusion of an individual, the boundary between the "normal" and the "outlaw."

Material Apparatus

1. The contacts indicate how agents adopting the same convention are interacting, if they often meet each other, on occasions regular or specific.

2. The technology indicates whether the link between individual and the convention is made via technical media, and whether this technique is a substitute for a human being, and therefore for the capacity to interpret rules.
3. The negotiation examines the degree of tolerance, enabling the interpretation of the rules without undermining, challenging, or calling into question the convention.

Complexity. With regard to the *complexity* of a convention, Gomez argues that, because a convention is an information and communication system, it is possible to apply the principles of general system theory (Gomez, 2006, p. 226). Thus, the more a convention provides a lot of different signals to the adopters and does not repeat them, the more complex it is. In this case, adopters have little room for interpretation. Conversely, a convention is slightly complex when it provides few rules but repeats them often. In the latter case, adopters have a lot of room for interpretation.

Linking morphology and complexity leads one to analyze a convention using the principle of *coherence* of a convention: Two elements of the morphology are *coherent* if they contribute to complexity in the same way. They are said to be *dissonant* if each of them has an opposite impact on complexity.

The Dynamic of Evolution of Conventions. This enables us to analyze the *dynamic of evolution* of conventions (Gomez & Jones, 2000): Depending on the degree of coherence, a convention can be more or less convincing for the adopters or potential adopters and therefore lead to confirmation, modification, or disappearance of the governing system (convention—individual agents). The governing system acts on individual agents and is acted on by them. Who governs the governing system? This system is a self-organizing and self-regulated system as behaviors and conventions (systems of rules) mutually and recursively interact according to the agents' shared conviction. Conventions are *dynamically stable patterns.* They evolve, modify themselves, or disappear according to the way the individual adopters change their behaviors over time.

In this evolution, both internal and external dynamics have to be considered. Internal dynamics is concerned with the process or routinization of behaviors (Dionysiou & Tsoukas, 2013; Salvato & Rerup, 2011). A convention provides individuals with conformist behaviors, fruit of a conformist calculation, i.e., which behaviors are considered "normal." The more individual routine is coherent with the screen of symbols, the higher the number of other individuals who will adopt the routine. Thus an individual routine becomes a collective routine, a routinized behavior, a non-justified rule for the adopters, and a "normal" behavior integrated in the rational void. However, conventions are never isolated, and alternative existing conventions lead to competition between the conventions, and to considering external dynamics. A convention gains adopters as it provides a better way, i.e., more coherent signals from the screen of symbols, to cope with uncertainty. If a competing convention is perceived, via signals received from the screen of symbols, to be better, i.e., bringing more coherence between individual and collective rationalization in addressing uncertainty, then the individual will shift. On this basis, a convention can resist, providing more coherent signals, adapt, modifying the screen of symbols and structure to gain more coherence and survive, or collapse, the adopters moving to a more convincing convention.

The Impact on Management. How can the individual action of an adopter change a convention? No individual action can alone change a convention as a whole; however, they are interdependent. Adopters of convention refer to the screen of symbols and thus never question the rational void. Facing dissonance as perceived by nonconformists, the adopters can react by reinforcing the screen of symbols either by persuasion, altering the signals to decrease the dissonance perceived by nonconformists; or by violence, rejecting the nonconformists and the dissonant signal. In both cases, adopters tend to protect the convention from doubt (mistrust) and from questioning the rational void. From the point of view of the individuals facing dissonance, they can accept it, move to a more "attractive" convention, or act to improve the coherence between the signals of the screen of symbols. The latter can be done through local and situational action on the screen of symbols. This discussion shows the role limit of any organizational and managerial action. Gomez & Jones (2000, p. 705) make it clear:

> Managers are not planners and decision makers applying a supposedly pure rationality, as they are always included in a social environment which gives both sense and limits to their rationality. They do not choose to act in one convention over another, but rather, as individuals, to escape the inhibiting effect of uncertainty. Once again, for any individual, the fact that the diversity of conventions allows some room for doubt and ambiguity is paradoxically the fact which gives them some freedom for action.

"Management" supposes a volunteer action on the conventions, i.e., on statement, material apparatus, complexity, and coherence/dissonance, can modify the conviction of the adopters toward the convention and can therefore support or change the governing system by acting on the system of rules.

4. To Not Conclude . . . Ω and A

The project management field is metaphorically like the Broceliande Forest: full of potentially highly powerful magic but tough to find his/her way. . . .

In this short text, I wanted to offer that, to unveil its full potential, the project management field needed to expand its horizons, from the classical approaches well suited when stakeholders are sharing some unitary perspective to a broader integrative view recognizing that, in the real world, stakeholders may have pluralistic perspectives. This extension involves a move to an integrative onto-epistemology, balancing on the one hand modernism and post-modernism (Boisot & McKelvey, 2010), and on the other hand, acting and knowledge creation and transfer (Bredillet et al., 2013a, 2013b). To illustrate this integrative lens, I propose that conventions theory may provide a robust theoretical background to the development and content of any framework (e.g., standard) aimed at addressing the challenge of value(s) creation in complex, ambiguous, and uncertain environments and situations.

This leads me to outline a general theory of "standards" as part of these (not) concluding comments. The way of conceptualizing "universals" or "general theory" has to be made clear. According to Eikeland (2008, p. 25), three kinds of traditions can be considered: (1) covering laws (deductive nomological or hypothetico-deductive model), (2) statistical generalizations, and (3) standards. Here, standards can be defined as "'fixed points' or 'ideals' for practitioners within certain areas, saying something about what it means to perform a certain kind of activity competently or, according to certain quality" (p. 26). The meaning does not include standards understood as just average norms, arbitrary or imposed by external bodies (e.g., Brunsson et al., 2000). Here, such standards are neither qualitatively nor quantitatively influenced by any counter-facts. Standards are made by the success of virtuoso practitioners, and they "change when someone finds a better way of doing, making or using something." The key characteristics of such standards are that "not everybody should or could realize them equally or fully . . . their non-arbitrary character, their immanence as patterns to practice, and "ways-of-doing-things," and their practical inevitability in human life as either implicit or explicit, vague or more exact standards of measurement, as standards of validity of excellence" (p. 26). Contrary to arbitrary standards, which can be conventional, unnecessary, or enforced, non-arbitrary standards are necessary as they express an existential necessity that is what it means to be or to do something. Such standards are to be observed practically *from within the practice,* and they are impossible to observe just from outside, by perception. The position of the "observer" is thus quite different in these three traditions. In the case of "standards," the observer is the practitioner, the native, dealing with things and theorizing his or her own practice, and there is no dichotomy between practice and theory (Eikeland, 2008, p. 27).

Finally, I suggest that organizations and professional bodies would get some benefits by being more conscious of the assumptions and of the dynamic and implications at stake underlying project management practice and development and design of routinized knowledge (e.g., standards).

I hope to have contributed, however humbly, to a better perception and understanding of this fascinating field Be-Have! if not bee-hive . . . (Marx, 1965), to a better understanding of the project management field and demonstrated that it, as an integrative field—the place of the mirror between past and future, analysis and foresight, logic and paradigm—offers unique characteristics. The main one is probably to contribute to transform reality into ideality!

Ordo ab chaos.

References

Anbari, F., Bredillet, C., & Turner, J. R. (2008, August). Exploring research in project management: Nine schools of project management thought. *Best Papers Proceedings, Academy of Management Conference,* Anaheim, CA.

Andriani, P., & McKelvey, B. (2009). From Gaussian to Paretian thinking: Causes and implications of power laws in organizations. *Organization Science, 20*(6), 1053–1071.

Argyris, C., & Schön, D. (1978). *Organizational learning: A theory of action perspective.* Reading, MA: Addison-Wesley.

Artto, K. A., Martinsuo, M., Gemünden, H. G., & Murtoaro, J. (2009). Foundations of program management: A bibliometric view. *International Journal of Project Management, 27,* 1–18.

Ashby, W. (1958). Requisite variety and implications for control of complex systems. *Cybernetica, 1,* 83–99.

Atkinson, R. (2006). Guest editorial: Excellence in teaching and learning for project management. *International Journal of Project Management, 24,* 185–186.

Audet, M. (1986). Le procès des connaissances de l'administration. In M. Audet & J. L. Malouin (Eds.), *La production des connaissances de l'administration (The indictment of knowledge in public administration)* (pp. 23–56). Québec: Les Presses de l'Université Laval.

Bakker, R. M. (2010). Taking stock of temporary organizational forms: A systematic review and research agenda. *International Journal of Management Reviews, 12*(4), 466–486.

Bernstein, R. J. (1983). *Beyond objectivism and relativism: Science, hermeneutics, and praxis.* Philadelphia, PA: University of Pennsylvania Press.

Boisot, M. H. (1998). *Knowledge assets: Securing competitive advantage in the information economy.* New York: Oxford University Press.

Boisot, M., & McKelvey, B. (2010). Integrating modernist and postmodernist perspectives on organizations: A complexity science bridge. *Academy of Management Review, 35*(3), 415–433.

Boltanski, L., & Chiapello, E. (1999). Le nouvel esprit du capitalisme. Paris: Gallimard.

Boltanski, L., & Chiapello, E. (2005). *The new spirit of capitalism.* New York: Verso.

Boltanski, L., & Thévenot, L. (1991). *De la justification. Les économies de la grandeur.* Paris: Gallimard.

Boltanski, L., & Thévenot, L. (2006). *On justification: Economies of worth.* Princeton, NJ: Princeton University Press.

Boutinet, J.-P. (1996). *Anthropologie du projet (Anthropology of project).* Paris: PUF.

Boutinet, J.-P. (1997). Tensions et paradoxes dans le management de projet (Tensions and paradoxes in the management of project). *Les cahiers de l'actif,* pp. 266–267.

Bredillet, C. (2003). Genesis and role of standards: Theoretical foundations and socio-economical model for the construction and use of standards. *International Journal of Project Management, 21*(6), 463–470.

Bredillet, C. N. (2004). Beyond the positivist mirror: Towards a project management "gnosis." *Proceedings of IRNOP VI,* Turku, Finland.

Bredillet, C. N. (2007). From the editor (series). *Project Management Journal, 38*(2–4).

Bredillet, C. (2010). Blowing hot and cold on project management. *Project Management Journal, 41*(3), 4–20.

Bredillet, C. N., Tywoniak, S., Hatcher, C., & Dwivedula, R. (2013a). What is a good project manager? Reconceptualizing the "do": An Aristotelian perspective. *11th edition of International Research Network on Organizing by Projects (IRNOP) Conference, "Innovative Approaches in Project Management Research,"* Oslo, Norway, June 17–19.

Bredillet, C. N., Tywoniak, S., & Hatcher, C. (2013b). Acting and knowing in temporary and project-based organizing: Turning from the practice world to a liberation praxeology? *European Academy of Management (EURAM) Conference, "Democratising Management,"* Istanbul, Turkey, June 26–29.

Bredillet, C. N., Tywoniak, S., & Hatcher, C. (2013c). Prolegomenon to the study of the concepts of maturity and maturity model: From black horse to white knight? *European Group for Organizational Studies (EGOS) Conference, "Bridging Continents, Cultures and Worldviews,"* Montréal, Canada, July 4–6.

Brunsson, N., Jacobson, B., and Associates (2000). *A world of standards.* Oxford, UK: Oxford University Press.

Cicmil, S., Williams, T., Thomas, J., & Hodgson, D. (2006). Rethinking project management: Researching the actuality of projects. *International Journal of Project Management, 24,* 675–686.

Comte, A. (1855). *Positive philosophy.* New York: Calvin Blanchard.

Cooke-Davies, T., Cicmil, S., Crawford, L., & Richardson, K. (2007). We're not in Kansas anymore Toto: Mapping the strange landscape of complexity theory, and its relationship to project management. *Project Management Journal, 8*(2), 50–61.

Crawford, L. H. (2007). Developing individual competence. In J. R. Turner (Ed.), *The Gower handbook of project management* (4th ed., pp. 677–694). Aldershot, UK: Gower.

Declerck, R. P., Debourse, J-P., & Declerck, J. C. (1997). *Le management stratégique: Contrôle de l'irréversibilité (Strategic management: The control of irreversibility).* Lille, France: Les éditions ESC Lille.

Dionysiou, D. D., & Tsoukas, H. (2013). Understanding the (re)creation of routines from within: A symbolic interactionist perspective. *Academy of Management Journal, 38*(2), 181–205.

Dogan, M. (2001). Paradigms in the social sciences. *International Encyclopedia of* *the Social & Behavioral Sciences* (pp. 11023–11027). http://dx.doi.org/10.1016/B0-08-043076-7/00782-8.

Eikeland, O. (2008). *The ways of Aristotle: Aristotelian phronesis. Aristotelian philosophy of dialogue, and action research.* Bern: Peter Lang.

Gareis, R., & Huemann, M. (2007). Maturity models for the project oriented company. In J. R. Turner (Ed.), *The Gower handbook of project management* (4th ed., pp. 183–208). Aldershot, UK: Gower.

Gentner, D. (1983). Structure-mapping: A theoretical framework for analogy. *Cognitive Science, 7*(2), 155–170.

Gomez, P.-Y. (2006). Informations et conventions: Le cadre du modèle général. *Revue Française de Gestion, 32*(160), 217–240.

Gomez, P. Y., & Jones, B. C. (2000). Conventions: An interpretation of deep structure in organizations. *Organization Science, 11,* 696–708.

Guénon, R. (1986). *Initiation et réalisation spirituelle (Initiation and spiritual fulfillment).* Paris: Editions Traditionnelles.

Hacking, I. (2002). Inaugural lecture: Chair of Philosophy and History of Scientific Concepts at the College de France, 16 January 2001. *Economy and Society, 31*(1), 1–14.

Hodgson, D., & Cicmil, S. (2006). *Making projects critical.* Basingstoke, UK: Palgrave MacMillan.

Houde, J. (2007). Analogically situated experiences: Creating insight through novel contexts. *Academy of Management Learning and Education, 6,* 321–331.

Jackson, M. C. (2003). *Systems thinking: Creative holism for managers.* Chichester, UK: Wiley.

Jolivet, F. (2003). *Manager l'entreprise par projets: les métarègles du management par projet.* Paris: EMS.

Kuhn, T. (1970). *The structure of scientific revolutions.* Chicago: University of Chicago Press.

Kurtz, C. F., & Snowden, D. J. (2003). The new dynamics of strategy: Sense-making in a complex and complicated world. *IBM Systems Journal, 42,* 462–483.

Kwak, Y. H., & Anbari, F. T. (2008). *Impact on project management of allied disciplines: Trends and future of project management practices and research.* Newtown Square, PA: Project Management Institute.

Le Moigne, J.-L. (1990). *La modélisation des systèmes complexes.* Paris: Dunod.

Le Moigne, J.-L. (1995). *Les épistémologies constructivistes (Constructivist epistemologies).* Paris: PUF.

Le Moigne, J.-L. (2003). *Le constructivisme—Tome 3 modéliser pour comprendre (Constructivism: Volume 3 modeling to understand).* Paris: Ed L'Harmattan, Coll. Ingenium.

Levi-Strauss, C. (1971). *The elementary structures of kinship.* Boston: Beacon Press.

Levi-Strauss, C. (1974). *Structural anthropology.* New York: HarperCollins.

Leybourne, S. A. (2007). The changing bias of project management research: A consideration of the literatures and an application of extant theory. *Project Management Journal, 38*(1), 61–73.

Long, C. P. (2002). The ontological reappropriation of phronesis. *Continental Philosophy Review, 35*(1), 35–60.

March, J., & Simon, H. (1958). *Organizations.* Chichester, UK: Wiley.

Marx, K. (1965). *Oeuvres, economie.* Paris: Gallimard NRF.

Maylor, H. (Guest Ed.). (2006). Rethinking project management. *International Journal of Project Management (Special Issue), 24*(8).

Menger, C. (L. Schneider, Ed.; F. J. Nock, Trans.). (1985). *Investigations into the method of the social sciences with special reference to economics.* New York: New York University Press.

Meredith, J. R. (2002). Developing project management theory for managerial application: The view of a research journal's editor. *Proceedings of PMI Research Conference 2002: Frontiers of Project Management Research and Application* [CD], Seattle, WA. Newtown Square, PA: Project Management Institute.

Morin, E. (1985). *La Méthode T 3, La connaissance de la connaissance.* Paris: Seuil Opus.

Morris, P. W. G. (1997). *The management of projects.* London: Thomas Telford.

Morris, P. W. G. (2013). *Reconstructing project management.* Chichester, UK: Wiley.

Polanyi, M. (1958). *Personal knowledge.* Chicago: University of Chicago Press.

Richardson, K. A. (2005). To be or not to be? That is (not) the question: Complexity theory and the need for critical thinking. In K. Richardson (Ed.), *Managing organizational complexity: Philosophy, theory, and application* (pp. 21–46). Greenwich, CT: IAP.

Salvato, C., & Rerup, C. (2011). Beyond collective entities: Multilevel research on organizational routines and capabilities. *Journal of Management, 37*(2), 468–490.

Sauer, C., & Reich, B. H. (2007). Guest editorial: What do we want from a theory of project management? A response to Rodney Turner. *International Journal of Project Management, 25,* 1–2.

Searle, J. R. (1997). *The mystery of consciousness.* New York: Granta.

Slack, N., Chambers, S., & Johnston, R. (2006). *Operations management.* London: Financial Times/Prentice-Hall.

Söderlund, J. (2011). Pluralism in project management: Navigating the crossroads of specialization and fragmentation. *International Journal of Management Reviews, 13,* 153–176.

Söderlund, J., & Geraldi, J. (2012). Classics in project management: Revisiting the past, creating the future. *International Journal of Managing Projects in Business, 5*(4), 559–577.

Stacey, R. D. (2010). *Complexity and organizational reality: Uncertainty and the need to rethink management after the collapse of investment capitalism* (2nd ed.). Milton Park, UK: Routledge.

Stacey, R. D. (2012). *Tools and techniques of leadership and management: Meeting the challenge of complexity.* Milton Park, UK: Routledge.

Turner, J. R. (2006). Editorial: Towards a theory of project management: The nature of the project. *International Journal of Project Management, 24,* 1–3.

Turner, J. R, Huemann, M., Anbari, F. T., & Bredillet, C. N. (2010). *Perspectives on Projects.* London and New York: Routledge.

Umpleby, S., & Anbari, F. T. (2004). Strengthening the global university system and enhancing projects management education. *Review of Business Research, 4,* 237–243.
Vico, G. B. (1708). *La Méthode des études de notre temps (The method of studies in our time).* Presentation & traduction par PONS Alain, 1981 (texte de 1708). Paris: Grasset.
Von Mises, L. (1976). *Epistemological problems of economics.* New York: New York University Press.
Von Mises, L. (1981). Praxeology. *The Freeman: Ideas on Liberty, 31,* 515–576.
Walker, D. H. T., Cicmil, S., Thomas, J., Anbari, F. T., & Bredillet, C. (2008). Collaborative academic/practitioner research in project management: Theory and models. *International Journal of Managing Projects in Business, 1*(1), 17–32.
Williams, T. (2002). *Modelling complex projects.* Chichester, UK: Wiley.
Winter, M., Smith, C., Morris, P. W. G., & Cicmil, S. (2006). Directions for future research in project management: The main findings of a UK government funded research network. *International Journal of Project Management, 24,* 638–649.

Part One

Polar Expeditions

Chapter 1

A Polar Expedition Project and Project Management*

Gilles Garel and Pascal Lièvre

This chapter pursues two objectives. First, it summarizes a project in which the authors participated, from the design stage to the final debriefing. The authors are, thus, both researchers and actors in the project. The project was a 2007 polar expedition by sea kayak along a section of the coast of Greenland. The expedition brought together four individuals who shared two kayaks and traveled in total autonomy over a distance of 150 km. Greenland, one of the world's largest islands, is located northeast of Canada, spanning 60° to 88° latitude north from its southernmost tip to its northern extremity. This chapter refers to the undertaking as a "polar expedition project" with "exploration and discovery" objectives. It emphasizes that this project represents one among many possible forms of polar expeditions (Lièvre, Récopé, & Rix, 2003). The expedition was a success, in the sense that the team not only returned with a high degree of satisfaction, but also experienced satisfaction throughout the project's realization. Although the project was a success, it encountered critical events along the way. The project summary draws on concepts developed by Bruno Latour (2005). These concepts are the basis for the descriptions of forms

* The authors extend their warmest thanks to Pascal Croset for his pivotal contributions to the expedition and subsequent analyses.

of socio-technical integration observed in the course of the project, which ultimately led the authors to tackle the question of how collective action was made possible. The summary constitutes the chapter's first objective. The second is to use the specific experience of the project to extrapolate conclusions for project management more broadly, including conclusions on team formation and the relationship between preparation and action, as well as to derive cognitive insights into knowledge management. The analysis of a specific experience can contribute to broader knowledge. By the same token, the study of expedition projects can contribute to project management research.

From a methodological standpoint, the use of a single case is justified, according to Yin (1981), when representing extreme situations—Greenland is extreme, both in its remoteness relative to other project locations cited in the relevant literature and in its environmental conditions—examining situations that had not been accessible previously to the research community (to the authors' knowledge, no other management researchers have organized and analyzed this type of project) or tested a theory. A deductive researcher works according to a logic that must either substantiate or contradict hypotheses derived from theory. A researcher's relationship with fieldwork is often indirect and mediated, by statistical instruments, for example. However, a project can be studied from within as well. An inductive researcher observes facts in practice in order to elaborate a theory progressively. This was our course of action. We embarked on an expedition—and we became, for this chapter, researchers.

In what way is a polar expedition a project? A degree of ambiguity undoubtedly surrounds its output. The development of new goods or services would surely constitute a relevant object of analysis for project management researchers. But what of a sea kayak expedition along the coast of Greenland? What was this project's deliverable? Both the intangible character of the output and the absence of a research dimension to the project (the team did not carry out any experiments *in situ* and pursued no other objective than returning to base camp) suggest that it was "recreational" in character. Yet is it possible to view the expedition members' memories, the "good times" spent together, as recognizable output? In other words, was this a "real" project? A polar expedition certainly does incorporate all the characteristics of a project, from the initial stages of preparation, through the expedition proper, to post-project capitalization. Conceptually, such an expedition exhibits all the attributes of a project as Midler (1996) defines the term, as Garel and Lièvre (2010) have shown, and as this book's conclusion reiterates.

The first section of the present chapter structures its discussion of the project around Latour's (2005) notion of controversy. The second section explores the conceptual and managerial insights deriving from the project.

1.1. The Expedition Project in the Light of Bruno Latour's Thought

The expedition began as a project among friends, without any thought to potential research goals. It was initially no more than a desire to travel together, to experience the polar world, in just a pair of sea kayaks. The plan included no major undertaking: The aim was a basic, quintessential sea kayaking excursion that would cover 150 km in total autonomy along the coast of Greenland from Ilulissat to Port Victor. The excursion had no other goals than to gain personal knowledge as an autonomous team completing a project, from the first embryonic idea through its realization. And in the aftermath of this short adventure (one year of preparation, 15 days of kayaking, financial closure and project review three months later) the team realized a "unanimous" and "strong" sense of satisfaction. In short, we aimed at a personal project for pleasure. Yet we quickly came to ask ourselves why we had felt such a high degree of satisfaction with this project. Perhaps, we wondered, the appreciation we felt for the project's qualities signaled that it could serve as an exemplary case study or, at the very least, that its examination could yield managerial, and perhaps personal, insight. In describing the expedition, we rely on the conceptual framework developed by Bruno Latour*: Actors, objects, places, and controversies are the components through which we examine combinations of human and nonhuman elements to shed light on the conditions that make a specific collective initiative possible.

1.1.1. Actors and Objects

Actors†

There were four principal actors, as well as one important secondary one. They were Paul, the perfectionist sailor; Pierre, the philosophical handyman; Joëlle, the outdoors sportswoman; and Philippe, the polar adventurer. The group of actors also included Bruno, the travel agent, who played an essential role, since it was he who was commissioned to arrange our flight tickets, boat transportation to base camp, and all equipment necessary for sea kayak travel.

* For a more in-depth treatment, please see Garel and Lièvre (2010).

† Names of expeditions as well as individuals have been changed for the purpose of confidentiality.

Objects and Places

The project involved close to a dozen nonhuman elements, physical objects, skills, management tools, and places, each playing a significant role over the course of the expedition and whose absence would have considerably affected the development of the collective initiative. During the expedition, we needed to make difficult choices.

Let us begin with our means of transportation—the Nautiraid—a collapsible softshell sea kayak originally developed for use by special military units. The particularities of the craft include its weight (40 kg), volume (equivalent to two large backpacks), and the fact that it can be quickly assembled and disassembled at any time. Philippe had long wished to use this type of craft, which makes it possible to carry out long excursions in polar regions during the summer, when climate conditions make it impossible to trek using skis and a pulk (i.e., a supply sled). Such summertime kayak excursions complement springtime treks for many polar expedition aficionados. Philippe had purchased a used two-seat Nautiraid brand kayak in the Grand Raid series. The model measured 5.30 m in length and could carry up to 350 kg of weight. Although this model is less efficient in terms of speed (it is capable of only half the speed of hardshell kayaks), it is highly reliable and easily repaired. It is also possible to mount a sail on the craft. Vessels of this type have successfully completed transatlantic crossings.

A second object playing an important role was treaded footwear. The black rubber boots we selected had deeply treaded soles that were adapted to travel over slippery, unstable ground; they are otherwise known as "caving" boots. At first, a significant discussion emerged among the team members as to the type of footwear to bring on the expedition. Our common backgrounds and experienced team members' opinions led us toward this type of foot protection; this proved a happy choice in the course of the expedition. The imperative was to keep our feet dry while embarking and disembarking from the kayaks, but without impairing the ability to walk safely over muddy terrain and wet rocks. In short, professional speleology boots were the ideal solution.

The style of drysuit to choose was also a matter of extensive discussion, in addition to providing moments of lighthearted diversion during fitting sessions with vendors. We finally opted for GoreTex suits, which combined good waterproofing with good aeration.

Philippe and Joëlle's family home served as headquarters for our expedition preparations. Our management tools included an equipment checklist, a list of necessary foodstuffs, and an expedition plan compiled using Excel software, which allowed us to make use of data collected in various situations over a number of years. A journal kept by a guide during a previous expedition, whose trajectory closely resembled ours, proved an invaluable source of information during the preparation stage and throughout the expedition's duration.

1.1.2. *Controversies*

We identified two important controversies that occurred during the course of the expedition. The first arose during the preliminary stages of project construction. The second occurred in the field, immediately following a crisis situation for the team, and involved choosing one of diverse scenarios for the journey toward our return flight.

Controversy #1: Project Construction

Philippe proposed that the expedition be a sea kayaking trip in the gulf of the Saint Lawrence River, where we could navigate alongside whales, which he felt would provide a good balance between his lack of experience in a sea kayak and the other three team members' lack of experience in polar regions. Aware that Philippe had previously driven along the banks of the Saint Lawrence and had paddled on the waterway on that occasion, Paul favored exploring more northern regions in order to make full use of Philippe's polar experience and in order to attempt something he would not have tried without the rest of the group. After reflection, Philippe proposed Greenland. The solution of a Greenland expedition satisfied the desires of both: for Philippe and Joëlle it would provide the opportunity to acquaint themselves with sea kayaking, for Paul and Pierre it would be a voyage of discovery to the far north. In addition, Greenland is easily accessible and provides all necessary logistics on site, and, what is more, Philippe had carried out expeditions in the region on three previous occasions.

Controversy #2: Return Scenarios

Eight days into the expedition, the team was caught in a swelling sea, when wind speed increased dramatically in the course of an hour. We found ourselves in a perilous situation, 2 km from shore, where we were running a high risk of capsizing into water no warmer than 2–5°C. We fought the waves for an hour before coming noticeably closer to shore. Running aground on the beach, we quickly sheltered the boats from the wind and the rolling waves. The sea was high. Wind gusts reached 70–80 km/h, while waves crested at

80–100 cm. The team searched for a place to camp, finding a suitable location in the proximity of a river. Philippe was resigned to dispense with traveling by kayak for the remainder of the expedition if the weather conditions did not improve. He gave serious consideration to building sleds and returning on foot through the valleys. Tension within the group was high. Paul felt that a trek on foot was a bad idea; he saw no reason why we should not return with the kayaks. Philippe considered yet another possibility: Perhaps a boat could pick us up and bring us to the return point. Pierre was uncomfortable with the idea of a "supervised" return, which would invalidate the project's autonomy. We knew also that another expedition was nearby and were to leave on Friday evening. We considered leaving the kayaks with them and making our return on foot. Philippe stated his case resolutely, map in hand: An autonomous return trip on foot was possible, and he was certain he could arrive at the return point by Friday evening. The following day, Wednesday, August 1, the weather remained unchanged; and although the meteorological prognosis called for a slight improvement, Philippe remained skeptical. Paul and Pierre went for a walk; on their return, the team decided to visit the "birds' cliff" together. Once there, we encountered the other expedition group and discussed the weather conditions with their guide. In confidence, he told us that, were he alone, he would brave the waters by kayak; but having a group under his supervision, he preferred to wait out the stormy weather. Around 5 p.m. the wind began to die down. Paul was ready to set off. Philippe did not agree, reasoning that we needed three hours to break camp, in addition to the time needed to eat and rest. He suggested waiting until the next morning to decide on a course of action. The next morning, the wind had abated. We took to the kayaks once again for a full day of paddling that would cover 40 km. We arrived at camp at 10 p.m., exhausted.

1.2. The Expedition Experience and Insights for Project Management

In this section, we examine certain characteristics of our expedition project that intersect with themes analyzed in project management literature and constitute factors of learning for future expedition projects. We focus on three elements.

1.2.1. Team Formation: Prototyping Potential Situations and Managing Expectations

Mutual affinity, trust, and respect were already present among three of the four expedition members, who knew each other before the project began, although they had never undertaken an expedition together. The fourth team member, Pierre, was the most recent addition to the group, through his friendship with Paul, who was, thus, the only actor familiar with everyone in the group from the outset of the project. In a development typical of project management processes, Paul paired with Pierre. Camaraderie notwithstanding, bringing together a group of friends is not the equivalent of constituting an expedition team. The latter required us to ensure that the group had the aptitudes necessary to operate as a cohesive team in difficult, perhaps dangerous, conditions; it was necessary, as well, to asses every member's capacity to persevere with the project, since, once we had set off, no one could withdraw until all of us returned to base camp*; and, finally, it was imperative that all members possess requisite physical capabilities.

One of this project's specific factors of interest for the authors was their own contribution to it, which had been thorough and continuous, from inception to conclusion. The preparation phase led to the expedition, that is, to the active phase of the project, which was irreversible or which, at the very least, made retreat extremely costly. The "upstream" phases of the project fulfilled specific functions that largely conditioned "downstream" performance. They included phases of training, testing, and prototyping, as well as phases of reconnaissance that later resulted in the localization of "outposts" from which expeditions could depart. The training phases served to assess the team's aptitudes for living together in close quarters and to test technical components (equipment, tools†). These phases were carried out in locations where conditions were representative of those projected for the expedition area. This stage consisted

* Retreat was, of course, always possible, as in the case of a clearly dangerous situation, for instance. The dimension we wish to underline here is the impossibility of leaving the project location once the expedition began. The team would remain a unit throughout the duration; there would be no breaks in the project to return home to one's private life only to re-engage in the project afterwards. Thus the team's management needed to incorporate such breaks into the project *in situ* (in the form of rest days, for example).

† We will not devote space here to the developments surrounding the "upstream" acquisition of resources; they have been amply covered in previous work on polar expedition logistics (Lièvre, 2003). It is nevertheless important to recall that the acquisition of resources calls on one to:

- Learn how to adjust one's orders of magnitude; realize that it is not the devil who is in the details but rather the very essence of the project; remember that resources that may seem inconsequential can become the fine, fragile threads on which an expedition may ultimately depend; anticipate which elements will prove vital, necessitating an expertise that is essential to the project's success;
- Above all, not be sparing in devoting necessary means to resources, because they will determine the comfort (ergo the morale) and safety of the expedition.

of important instances, during which the expedition members attempted to work out a partial depiction of future objects and processes based on existing knowledge. Situation modeling played a crucial role in these stages and had the potential to lead expedition members to devise new tools and innovative action mechanisms. The actors received assistance during this phase from partners and experts proficient in the use of such tools. Project prototyping took place in a "representative" environment (Thomke, 2003). Specifically, the experienced members of the team undertook camping trips in winter (March 2007) and in spring (June 2007), using the same equipment they planned to use in Greenland, in order to evaluate the three dimensions discussed above.

The process of recruiting members for the expedition and constituting the team was also a process of managing expectations. Indeed, the specific attention devoted to expectations management in finalizing the project was a fundamental element in establishing group cohesion. This process was carried out from the outset of the "upstream" phases, when other dimensions of the project remained open to modification. Philippe, the expedition leader, directed the expectations management phase in a manner that was progressive, collective, transparent, and nonmanipulative. For instance, Philippe had in mind an initial idea of a possible locale for the expedition. The first weekend of winter preparations consisted of conceptualizing the expedition's various dimensions, which remained entirely open at this stage, including the destination. In order to define this dimension, Philippe asked the other team members to express their expectations. He guided the collective search for a destination based on a progressive exchange of expectations, while maintaining the process within a precisely defined framework of parameters established from the outset (e.g., two kayaks and two weeks). Since Philippe was the team member with the most knowledge relevant to the expedition's scope of possibilities, he was best placed to guide the convergence of all members' expectations—on the condition that he, too, would respect the guidelines he had laid down. We can easily imagine a similar situation, where a group leader could have adopted a far more manipulative approach, concealing his own project, that is, the satisfaction of his own ambitions. It would have been quite straightforward for Philippe (due to the asymmetry of information and experience) to direct the team's choice of destination toward one of his own preference, without disclosing his intention to the team. The measure of success of the convergence/definition phase is the degree of dedication to the collective project that the process generates. Beyond practical aspects such as fitness training, the forthrightness and transparency of the decision-making process proved to be pivotal dimensions of team formation during our weekend preparation sessions.

Examining the motivations of a team's commitment to a project means also considering potential disparities between the members' declared motivations and what really motivates them. Indeed, to question the motivations of commitment to a project team is ultimately to take into account the alignment of motivations among the individuals concerned. Baron (1993) considers that an individual's performance in the course of a project is linked less to criteria such as personality profile, aptitudes, and skills than it is to the consistency of the individual's personal project with the professional project at hand, as well as what the individual believes can be gained by committing to the project. In this manner, the team formed around the expedition leader through processes of co-optation and *in situ* testing.

1.2.2. The Relationship between Preparation and Action: Adapting and Preparing to Improvise, Rather than Planning

The challenge of a project's "upstream" phases is to develop theories of action and to secure tangible resources. Yet the key insight we gained from the polar expedition was that it is imperative to adapt *in situ*, even when this means diverting from what has been planned. Most successful expeditions, in fact, detour from their planned routes and renegotiate their initial objectives during the course of events. The teleological vision of planning, in which project management consists of a set of actions designed to correct deviations from fixed objectives, was inadequate for a polar expedition such as ours, not least because meteorological forecasts were inaccessible and wind conditions highly localized. Glaciers at the head of fjords often cause or accelerate katabatic winds, which present dangers for heavily laden and not particularly maneuverable kayaks navigating glacial coastal waters. Prevalent concepts of cost control and earned value involve management tools that define targets *ex ante*, by which deviations are measured against established reference points as the tasks awaiting completion diminish over the course of a project. In such conditions, leading a project consists of steering it back toward its goals in case of deviation, while performance consists of respecting guidelines, whether they remain unchanged or have been redefined (Duncan, 1996).

A more adaptable approach to strategic planning consists of planning with a broader outlook that incorporates "just-in-case" components (Mintzberg, 1994). This approach has been termed "planned flexibility" (Verganti, 1999) or, inversely, flexible planning; however, in either case, it is no longer planning as it was once conceived. In fact, the Scandinavian school of project management has entirely replaced planning by the process of enactment, in which participants enact a project environment as part of a process of continuous learning (Weick, 1969).

The success of our expedition project was not, therefore, the result of adherence to "upstream" planning, but rather the consequence of *in situ* adaptation to conditions encountered in the here and now. The "upstream" phase, therefore, is not contingent on planning or the illusory mastery of the concomitant risks (e.g., meteorological forecasts), but depends instead on meticulous preparation for a range of possible situations. The expedition leader and the team must be able to adapt, revise, and improvise *in situ*. Improvisation, in fact, is the fruit of successful preparation. The adjustment of theory in the course of practice (Schön, 1983) is linked to a learning process that begins *before* the departure and continues *throughout* the expedition. The adopted theories were thus called into question by interactions among expedition members and by the physical environment (incidentally confirming the importance of the capacity to interpret the environment's faint signals). Heers (1981), for example, identified the same process at work in Christopher Columbus's voyage of discovery, noting the significant role that discussions between the explorer and scholars accompanying him to Hispaniola played in the development of his representations.

1.2.3. Knowledge and Ignorance: Knowledge in Action

Project management literature abounds in studies of skills, including those of team leaders and those of team members (Turner & Müller, 2005; Picq, 2005). Skills are contingent: The "upstream" and "downstream" phases of a project call on different capabilities and, moreover, capabilities will vary according to project characteristics. The Greenland expedition generated two particularly noteworthy cognitive insights.

Cognitive Gaps and *in Situ* Knowledge Production

Philippe's polar expedition experience was already significant when our project began to take shape, comprising close to a dozen such expeditions. On the other hand, he had never used a sea kayak or traveled to Greenland during the summer. Thus there was a cognitive gap relative to sea kayak use in Greenland, which, in the circumstances, none of the team members could remedy. Philippe looked forward to learning from the expedition: Mastering the sea kayak was one of his motivations. In the event, the only serious difficulties the team encountered were connected to navigating the kayaks at sea. In other words, the project could have failed due to this lack of knowledge. When in a situation, the key challenge is to make decisions that benefit the project. In our case, this meant decisions that did not endanger the team. But how could the expedition leader determine whether a particular trajectory steering the kayaks through a fjord was dangerous, without knowledge that had never been available to him previously? Such dilemmas link back to old debates on the illusion of an omniscient project leader (Midler, 1993). In practice, the leader must observe meta-rules (Jolivet & Navarre, 1993) of prudence; that is, he or she must identify areas of inexperience, *a priori*, and implement appropriate solutions before launching the irreversible phase, drawing on the expertise of other actors (local experts in particular), or be resigned to abandon certain elements *in situ*. While still in Greenland, we held a debriefing, in which we touched on mistakes in navigation decisions, and later engaged in extensive postexpedition discussions with sea kayaking specialists. These exchanges proved to be invaluable sources of knowledge, both during the expedition and for subsequent projects.

An unforeseen event during the expedition also became a source of specific *in situ* learning. One day, Paul and Pierre changed course in their kayak to approach a glacier by themselves. The two novices, who were supposed to remain within Philippe's field of vision, disappeared from view for nearly two hours. This manifestly hazardous situation, caused by the duo's initiative, can be traced back to a "black hole" within the project's framework. Indeed, Pierre and Paul had broken no project rules, whether formal or implicit. Yet the situation had created a risk that was difficult for the expedition leader to manage. Thus, a rule ("we do not separate, at least never without informing the others and only by mutual agreement") emerged *ex post,* immediately following a strongly worded reunion. The need for this rule was made obvious by the expedition leader's exasperated question to the two aspiring explorers: "And what will I say to your wives when you don't come back and I'm the one responsible for the project?" Thus the team produced a rule based on a specific situation that served as a learning event.

Generic Knowledge and Local Knowledge

Another project characteristic noteworthy in terms of management was the manifest difference between the dimensions of "generic knowledge" and "local knowledge." Philippe's knowledge at the outset was already considerable, especially in relation to the lack of knowledge of the two novice team members, the consequence of two decades of diverse expedition experience and a commitment to "reflexive practice." Nevertheless, *in situ*, during the expedition, local knowledge and generic knowledge became mutually complementary. To know how to cross a glacier is not necessarily to know how to cross this glacier, here and now. Familiarity with navigation by sail does not necessarily equal knowing how to navigate fierce, opposing currents in the Gulf of Morbihan. An expedition's success depends on local knowledge. In other words, it is *in situ* implementation, that is, precise and localized utilization, that determines a project's success. Generic knowledge is an essential resource in the production of situated, local knowledge. The capacity to interpret the environment and the faint signals it provides, as well as the capacity to assume one's ignorance ("in the here and now, I don't know or I know too little") are determinant for project success. The development of local innovations can provide solutions for cognitive gaps. The Greenland expedition team remained alert to innovative solutions that facilitated improvements to safety and comfort. Our kayaking expedition generated situated innovations by capitalizing on environmental conditions. This process was in operation, for instance, when we tested various sails fashioned from tarps, providing a welcome respite from rowing as gentle tailwinds carried us across a calm sea.

1.3. Conclusion

A polar expedition project has all the characteristics of an activity that is temporary, physical, combinatorial subject to uncertainty and exogenous factors, has an overarching goal, and consumes resources. It is therefore entirely legitimate to examine a polar expedition within the conceptual field of project management. This project, whose deliverable was neither a new product nor the construction of infrastructure, can be considered in terms of event-based project management. We applied an inductive (or, more precisely, abductive) qualitative methodology to a single case, in accordance with Yin (1981), from an internal perspective. The study is an in-depth examination of a polar expedition of exploration and discovery. The authors were also actors in the project. In our analysis, we have referred to the conceptual framework elaborated by Bruno Latour in order in order to guide our description of the development of the polar expedition project.

Three conclusions emerging from the analysis are particularly noteworthy in terms of their contribution to the managerial insights of project management literature.

The first concerns team formation and the crucial process of reconfiguration directed by the project leader in relation to the expectations of team members. The clarification of teammates' expectations and the leader's approach in reconfiguring those expectations are crucial. On the one hand, the expression of team members' expectations goes beyond the declarative dimension. It is in fact an exercise in uncovering deep currents that have marked actors' life paths. It is for this reason that the project leader proposed, from the very first meetings, that the team should undergo relatively difficult and engaging experiences during preparation activities in order to better grasp their expectations *in situ*. On the other hand, assessing the diverse expectations within the perspective of the project's configuration/reconfiguration process necessitates a deliberative mechanism that engages all actors and significant resources (directed *ex ante* by the project leader) in order to devise solutions *in situ*.

The second important conclusion relates to the ongoing revisions made to the initial project plan in the field, throughout the expedition. This process required the integration of uncertainties linked not only to weather conditions, but also to the team's constant need to feel out the terrain and proceed by trial and error. Throughout the project, from inception to realization, the group was involved in what Schön (1983) defined as an organizational learning situation. The project plan is a resource for action, but it must ensure adequate leeway for improvisation.

The third conclusion of note concerns knowledge management and, more precisely, the cognitive gaps identified *ex ante* and the transition from generic to local knowledge. The project leader and other team members were aware of each others' lacks of knowledge and skills in areas important to the project at hand. The lack of sea kayaking experience affected everyone in the team, while a lack of polar expedition experiences impacted only some members. These lacks or gaps were the cause of difficulties encountered during the expedition. The manner in which such deficits in knowledge can be identified must be addressed already at the preparation phase. Yet the inventory of these cognitive gaps always remains incomplete and the methods adopted to resolve them are never fully sufficient. It is therefore imperative to focus team formation on shared values, the capacity to listen to others, and mutual trust—elements that allow a group to face entirely unpredictable situations by mobilizing what Weick (1993) has termed "organizational resilience."

References

Baron, X. (1993). Les enjeux de la gestion des salariés travaillant dans les structures projet [The challenges of managing employees in projects]. *Gestion 2000, IX*(2), 201–214.

Croset, P. (2008). Dialogues praxéens [Praxéens dialogues]. Retrieved March 25, 2010, from http://www.praxeoconseil.fr/dialogues_praxeens.html

Duncan W. (1996). *A guide to the project management body of knowledge.* Philadelphia: Project Management Institute.

Ekstedt, E., Lundin, R. A., Söderholm, A., & Wirdenius, H. (1999). *Neo-industrial organising, renewal by action and knowledge formation in a project intensive economy.* London: Routledge.

Garel, G., & Lièvre, P. (2010). Polar expedition project and project management. *Project Management Journal, 41*(3), 21–31.

Hällgren, M. (2007). Beyond the point of no return: On the management of deviations. *International Journal of Project Management, 25*, 773–780.

Heers, J. (1981). *Christophe Colomb.* Paris: Hachette.

Jolivet, F., & Navarre, C. (1993, April). Grands projets, auto-organisation, méta-règles: Vers de nouvelles formes de management des grands projets [Large-scale projects, self-organization, meta-rules: Toward new forms of management of major projects]. *Gestion 2000,* pp. 191–200.

Kloppenborg, T., & Opfer, W. (2002). The current state of project management research: Trends, interpretation and predictions. *Project Management Journal, 33*(2), 5–18.

Koehn, N., Helms, E., & Mead, P. (2003, April). Leadership in crisis: Ernest Shackleton and the epic voyage of the *Endurance.* Harvard Business School Case Study (Prod. # 9-803-127).

Latour, B. (2005). *Re-assembling the social: An introduction to actor network theory.* Oxford, UK: Oxford University Press.

Lièvre, P. (2003). *La logistique des expéditions polaires à ski* [*Logistics of polar ski expeditions*]. Paris: GNGL Productions.

Lièvre, P., Récopé, M., & Rix, G. (2003). Finalités des expéditeurs polaires et principes d'organisation [Aims of polar expedition members and organizing principles]. In P. Lièvre (Ed.), *La logistique des expéditions polaires à ski* [*Logistics of polar ski expeditions*] (pp. 85–101). Paris: GNGL Productions.

Lièvre, P., & Rix-Lièvre, G. (2009). L'observatoire de l'organisant: Mode d'interprétation des matériaux qui en sont issus [Analysis of data from in vivo observation of organizing]. *Revue Internationale de Psychosociologie, XV*(1), 161–178.

Lundin, R. A., & Söderholm, A. (1995). A theory of the temporary organization. *Scandinavian Journal of Management, 11*, 437–455.

Midler, C. (1993). *L'auto qui n'existait pas; Management des projets et transformation de l'entreprise* [*The car that did not exist; Project management and business transformation*]. Paris: InterEditions.

Midler, C. (1996). Modèles gestionnaires et régulations économiques de la conception [Management models and economic regulation of design]. In G. De Terssac & E. Friedberg (Eds.), *Coopération et conception* [*Cooperation and design*] (pp. 63–85). Toulouse, France: Octares.

Mintzberg, H. (1994). *The rise and fall of strategic planning.* New York: Free Press.

Picq, T. (2005). *Manager une équipe projet* [*Managing a project team*]. Paris: Dunod.

Salembier, P., & Zacklad, M. (2007). *Annotations dans les documents pour l'action* [*Annotations in documents for practice*]. Paris: Hermès Lavoisier.

Schön, D. A. (1983). *The reflective practitioner: How professionals think in action.* New York: Basic Books.

Thomke, S. (2003). *Experimentation matters: Unlocking the potential of new technologies for innovation.* Boston, MA: Harvard Business School Press.

Turner, J. R., & Müller, R. (2005). The project manager's leadership style as a success factor on projects: A literature review. *Project Management Journal, 36*(2), 49–61.

Verganti, R. (1999). Planned flexibility: Linking anticipation and reaction in product development projects. *Journal of Production Innovation Management, 16,* 363–376.

Weick, K. E. (1969). *The social psychology of organizing.* Reading, MA: Addison-Wesley.

Weick, K. E. (1993). The collapse of sensemaking in organizations: The Mann Gulch disaster. *Administrative Science Quarterly, 38,* 628–652.

Yin, R. K. (1981). The case study crisis: Some answers. *Administrative Science Quarterly, 26*(1), 58–65.

Chapter 2

Ambidexterity as a Project Leader Competency: A Comparative Case Study of Two Polar Expeditions*

Monique Aubry and Pascal Lièvre

The chapter explores and clarifies the tensions placed on a project leader who is under the strain caused by different modes of action during a project. The aim is to describe these different modes and to understand opportunities for changing modes and each mode's inherent pertinence. We hypothesize that March's studies (1991) on ambidexterity in management science can help clarify these tensions, because March focused on the question of arbitration between exploitation mode and exploration mode in organizational learning processes. However, this view of ambidexterity fails to account for all the modes of action that a project leader can enter during a project, and we therefore suggest complementing it with the notion of ambidexterity put forward by Mintzberg (1994). We also apply James's (1907) principle of pragmatism to a comparative case study of events during two very different polar expeditions: one in the Arctic, called ARC, and the other in the Antarctic, called ANT.

* A longer version of this article was published in the *Project Management Journal* (Aubry & Lièvre, 2010).

We consider project management to be a particular class within a mode of organizing, since it involves managing the emergence of a new activity, which must be distinguished from a standard activity. This novelty aspect is more or less explicit. The introduction of novelty, and thus of the unknown, generates highly specific organizational problems (Garel, 2006). We agree with Midler (1994) that this type of organization must arbitrate a kind of tension between two factors that move in opposite directions: knowledge of a project and the degree of freedom involved in carrying it out. On the one hand, knowledge of a project progressively grows as it advances, and by the end, "everything" is known. On the other hand, the degree of project management freedom gradually diminishes; in other words, no room for maneuvering remains at the end of a project, when there is full knowledge of the project. This confronts us with a general problem related to organizational learning processes.

2.1. Suggested Analytical Framework

2.1.1. The Dual Nature of Project Ambidexterity

According to the Project Management Institute, "A project is a temporary endeavor undertaken to create a unique product, service, or result" (Project Management Institute, 2013, p. 3). Within a limited space and time, a project enlists novelty in contrast to routine or repetitive activities. Seen from this angle, a project seems geared largely toward an exploratory mode, while observation of how project management occurs within organizations seems much more closely related to an exploitative mode. When confronted with the need to place project management at their center, organizations have tended to rely increasingly on standardization (Bredillet, Yatim, & Ruiz, 2010).

We note considerable dissatisfaction with project management on theoretical and practical levels. Project success rates are unsatisfactory. Major public projects, like those of large organizations, often end in failure. Researchers blame traditional positivist principles inherited from Taylorism (Williams, 2005). One criticism often leveled at project management is the overrationalization of its planning. The logic of rationalization implies that the application of good practices and procedures leads to project success. According to this approach, failure results from a lack of rigor in applying these practices and procedures. The rationalization approach aims to eliminate or manage unplanned events to ensure that the project moves forward as planned. Otherwise, a project fails. Accordingly, developing a project plan involves using technical knowledge to structure and coordinate a series of activities at the best possible cost and within the shortest period of time. Once it is developed, a project leader then *implements* the plan.

Therefore, a project leader seems prone to a dual tension: the tension between exploitation and exploration modes and the tension between rationalization or planning and adaptation modes. We intend to clarify these different types of tension with the concept of ambidexterity developed by March (1991) and by Mintzberg (1994).

2.1.2. Ambidexterity According to March

March (1991) opened up a new line of research by placing the dilemma of exploitation versus exploration at the heart of his work on organizational learning (Farjaudon & Soulerot, 2008). Exploitation and exploration are no longer studied solely through the lens of strategy implementation but also through that of the knowledge possessed or to be acquired by the organization.

Based on these writings, some researchers see an incompatibility between exploitation and exploration strategies in the same organization over the same time and space horizon (Mom, Van Den Bosch, & Volberda, 2003). Others, on the contrary, demonstrate the necessity for an organization to use both of these modes, which they call its capacity for "ambidexterity" (Rivking & Siggelkow, 2003). From this perspective, various studies demonstrate the options available for combining these two forms of logic within an organization by identifying various types of ambidexterity. Three forms of ambidexterity involving a relatively discernible integration of exploitation and exploration stand out in the literature: structural, contextual, and network-based (Brion, Favre-Bonté, & Mothe, 2007; Garel & Rosier, 2008; Ney, Favre-Bonté, & Baret, 2008). Structural ambidexterity refers to organizations that separate their exploration activities in order to avoid the inertia often associated with exploitation activities and that hinders the emergence of innovation (Tushman & O'Reilly, 1996; Benner & Tushman, 2003). Contextual ambidexterity emphasizes an organization's capacity to reconfigure its organizational activities quickly (Birkinshaw & Gibson, 2004). In network ambidexterity, exploitation and exploration activities are divided among legally distinct entities that function as a network. Sometimes the focus of ambidexterity studies shifts from organizational design to the competences required to sustain innovative exploitation (which relies on existing competencies) and innovative exploration (which steps away from existing, centralized competencies (Chanal & Mothe, 2005). Although Garel and Rosier (2008) argue for further clarification of these notions, we propose to turn our attention to this notion of ambidexterity and examine it, not at the organizational level, but at the individual level of a project leader. This approach to ambidexterity raises the question: Is it a matter of using existing competences (exploitation mode) or, rather, to acquire new competencies (exploration mode)?

2.1.3. Ambidexterity According to Mintzberg

Mintzberg (1994) proposed another way of approaching the issue of ambidexterity based on the left brain/right brain perspective to address strategy, and this approach is based on the work of Physiology Nobel Prize winner Roger Sperry. All complex human activity combines these two modes of thought at the same time.

These two modes of thought can be combined in a highly productive manner but cannot be reduced to a single process. Any serious engineering design project requires both analysis and synthesis. We are familiar with Mintzberg's argument supporting the full use of these two forms of thought in guiding the strategic direction of organizations: plan with the left side, and manage with the right. This full use of these two forms of thought can be translated in terms of ambidexterity. Mintzberg (1994, p. 360) illustrated (see Figure 2.1) in a matrix the relations between rationality and learning at work during different strategies implemented through a given action. This matrix helps to shed light on the debate about modes of action at work in projects.

To pursue further our investigation of ambidexterity's dual nature in project leaders, in the sense of March and Mintzberg, we turn to the pragmatic method of William James (1907), which entails clarifying abstract notions based on

Was the achieved plan successful?	Was the intended plan achieved? *Yes*	Was the intended plan achieved? *No*
Yes	**A** Deliberate success (rationality)	**B** Emerging success (adaptation and learning)
No	**C** Failure of deliberate strategy (efficiency but not effectiveness)	**D** Complete failure (possibility of learning)

Figure 2.1 Rationality and/or learning during action.

concrete situations. To clarify our abstract notions of ambidexterity, we use actual polar expeditions as concrete situations.

2.2. Comparative Case Study of Two Polar Expeditions

2.2.1. Methodology

A polar expedition is approached as a full-blown project activity (Garel, 2003), from a deliberately generic angle. Essentially, we are working with case studies understood as intermediary theoretical situations within the meaning of David (2000) to allow us to establish permanent relations between revealed facts and the body of management science knowledge.

The Organization of a Polar Expedition

Our subject of interest is polar expeditions: from the first idea of the project and its implementation to the final financial closing. A polar expedition is a temporary organization arranged in project form. It emerges around a relatively specific objective such as making a crossing, reaching a summit or a mythical milestone, or pursuing a scientific or recreational purpose. We might say that the organization emerges from the moment that a project is discussed in a specific manner. A certain number of actors rally around an expedition's objective. Depending on the level of difficulty involved in an expedition, preparation time can take from six months to two years. During this time, a group seeks documentation, meets with experts, assigns tasks, develops the ideal plan for the expedition, purchases and tests equipment, provides individual and group training, and gathers the administrative and financial documents needed to depart on the expedition. An expedition itself occurs in the field over a specific period of time lasting from eight days to more than a year and starting and ending with a round-trip flight to and from an expedition location. An expedition ends not when its members return to their respective homes, but when its books are closed and commitments made to sponsors, scientific partners, or organizations that subsidized the expedition have been fulfilled. Depending on the aforementioned activities, an expeditionary organization may remain active for several months or several years after an expedition, in the strict sense, ends.

This type of project seems to have a few special features. The same team often remains in place throughout an entire project. It seems that the field implementation phase always escapes full planning, and that the right attitude for team members is not strict adherence to the plan but openness to constantly adapting the plan. A plan is treated as an overall resource for taking action in specific situations. At the same time, too little forethought and preparation can have serious future consequences during an expedition. Team members' acquired experience is an invaluable source of knowledge. Anything new must be tested before being implemented during an expedition.

In this particular case, we distinguish between two phases in the execution of a project: a design phase and an implementation phase. In polar expeditions, the design phase of an expedition begins with the idea of the project, and the implementation phase of the expedition begins with the journey to the actual expedition site.

Two Contrasting Polar Expeditions

This chapter is a comparative study of two very different expeditions to which we apply the same interpretive model to account for different operations that comprise these two phases: a design phase and an implementation phase. We try to identify the mode of action used by the project leader in each project phase: exploitation or exploration mode, planning or adaptation mode. We used two very different methods to gather data while pursuing the same objectives and exhibit strong methodical opportunism within the meaning of Girin (1990). Indeed, the very nature of the two expeditionary fields conditioned our investigation methods. On one hand, the personal nature of the ARC expedition implied that researchers would be involved ahead of the project and would monitor the expedition closely, since members of the expedition would not be creating a document that would tell their story. On the other hand, the high media profile of the ANT expedition made the presence and approach of researchers difficult, yet at the same time, the expedition had a daily log of its activities, broadcast live by the media.

The data from the ARC expedition was collected at the same time as the expedition unfolded by an "observation-participation" approach (refer to Chapter 5 and Rix-Lièvre & Lièvre, 2009). The data can be divided into two types: data from a researcher's logbook, the purpose of which was to report on events during the expedition from a group point of view; and data generated by discussions held by another researcher on how each member of the team experienced the expedition. The first researcher took an observation-participation approach. To compile the logbook, a variety of materials were used: daily notes, video recordings with commentary, and interviews conducted during and after the expedition. The second researcher approached the subject from an observation-participation angle that let him understand how the expedition unfolded in detail yet retain some distance from the collective experience, in order to gather more intimate accounts from the team members. It also included conversations held after the expedition about specific moments in relation to the objectives of the research conducted. The technique used was that of a *subjective re situ* (Rix & Biache, 2004), which involves prearranged video recording *in situ*.

The ANT expedition data was collected *a posteriori* and can be classified into two types. The first type of data includes writings by the expedition leader and crew members. The main source is the logbook, with entries made daily during the expedition. A powerful communications strategy was developed in order to maximize the visibility of the expedition and to focus public attention on climate change and the actions that must be taken by all to mitigate climate change. In total, more than 100 pages were analyzed. The second type of data, much less voluminous, includes the transcript of two one-hour interviews with the expedition leader. The objective of the first interview was to become better acquainted with the expedition leader, his history, values, and feelings. The second interview focused more specifically on preparation and planning activities.

2.2.2. The Two Expeditions*

The ARC Expedition: A Sporting Expedition by Ski

The ARC expedition took place in the Arctic. The objective was to make an unassisted crossing of the Arctic on skis. While the nature of the expedition implied some physical prowess, the objective was discovery. The team of four members was led by Joël, who took part in organizing and implementing the adventure. Joël and other team members made arrangements for the expedition in their spare time. This was a private expedition where each member financed his own way. The expedition took place over a period of two-and-a-half years, from the idea stage right up to the closing of accounts and the feedback on the experience. The idea developed over a year's time and materialized in the form of a project that would lead to a year's worth of preparations. The field phase lasted a little more than one month. The financial closing took place quickly in the month following the expedition's return to France (see Figure 2.2).

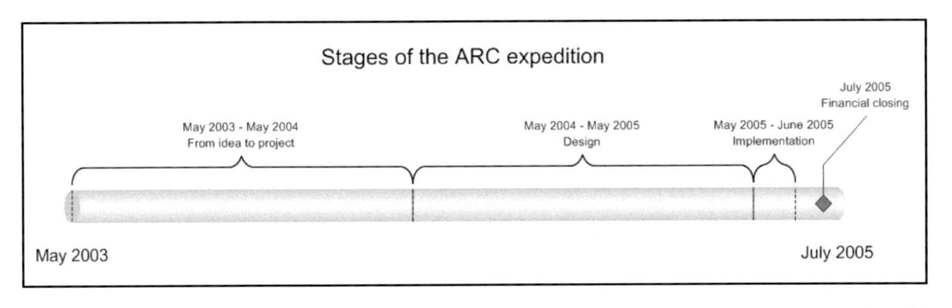

Figure 2.2 Stages of the ARC expedition.

* Names of expeditions as well as individuals have been changed for the purpose of confidentiality.

The ANT Expedition: A Large-Budget, Discontinuous, High-Media-Profile, Ship-Based Scientific Expedition

ANT was a ship-based expedition in the Antarctic. Its purpose was scientific. The project leader, Eric, was a scientific expedition professional and had invested everything he had in the project. This expedition was serious and entailed extensive advance preparation by many partners, specialists, and sponsors, as well as post-expedition activities, such as conferences, media coverage, and newspaper articles to meet sponsors' expectations. The team consisted of about ten members who would play different roles at different times during the course of the expedition. The ANT expedition covered a period of a little more than three years, excluding closing the books and publicity after the expedition team had returned. Preparations lasted two years, while the expedition strictly speaking lasted 15 months (see Figure 2.3).

After this brief presentation of the two expeditions, we now examine the management methods used in a critical situation encountered in each expedition to understand the ambidexterity of each expedition's project leader.

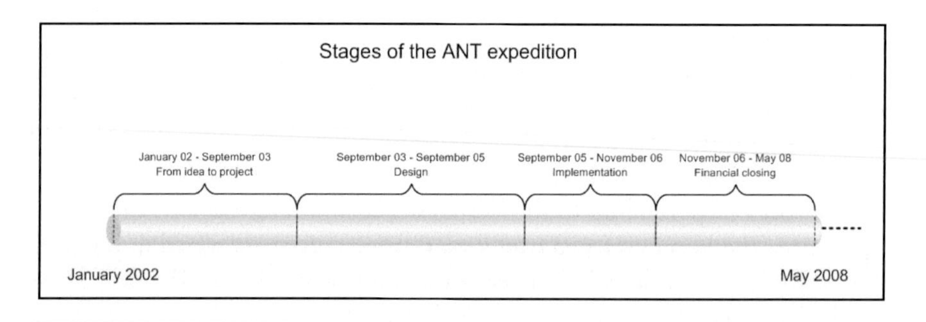

Figure 2.3 Stages of the ANT expedition.

2.3. Management Methods Used in a Critical Situation during Each Expedition

We examine a critical situation encountered in each expedition in an effort to make some initial conclusions in this investigation of ambidexterity. For the ARC expedition, the critical situation involved protection from bear attacks. For the ANT expedition, the critical situation had to do with mooring the ship in the Bay of Whales.

2.3.1. The ARC Expedition's Critical Situation: Protecting against Bear Attacks

At one meeting during the preparation phase, the discussion turned to protection against polar bears. Opinion was divided, but in any case, they had to bring a gun. It was a mandatory tool to bring along for this kind of expedition. A second question was raised about the possibility of bringing two guns, one for each tent. Joël, the expedition leader, decided it would be too dangerous and stated, "We just need one gun and I will carry it." There was concern about protecting against bears at night. Several options were mentioned, which included rotating watches and setting up a warning system around the camp. Joël settled the matter by suggesting that they bring dogs, which he had done on previous trips without being responsible for dogs. The others had never used dogs and were skeptical because the dogs could attract bears, too. Joël said that he and Gilles would take charge of the dogs during the expedition. Joël ruled the matter as settled and decided to buy two dogs on location.

Once at the location, Joël applied his strategy, though it was not without problems. On the second day of the expedition, the weather was good but the wind was strong. Making headway with the dogs was difficult; they had never done this kind of activity before—walking beside a skier while pulling a small sled for their food. There was a 300-meter slope to cross before reaching the pass, which posed a great physical challenge, especially with a sled weighing about 90 kilos. Joël left early that morning to mark a trail, and it was cold. One of his thumbs froze. He did not say much about it. By the early afternoon, the wind was blowing harder, and Joël had still not managed to warm up his thumb. The team grew worried. Pierre decided to set up camp immediately to allow Joël to warm up. The next day, Joël left as if nothing had happened. He went on ahead with a dog without saying a word. The others followed and then caught up with him. They arrived at a tricky crossing over an ice-covered river that required putting on crampons. The dogs were afraid; they kept slipping. Joël and Gilles, who were supposed to handle the dogs, were in the front. Pierre and his friend were behind with the dogs. They were making slow progress. They came to a crevasse in the ice, a gap of almost 3 meters. Pierre advanced slowly with a dog and slipped. To avoid hurting the animal, he let it go and sent it toward Joël, who was waiting at the bottom of the slope. Instead of rushing toward the dog, Joël figured, "He's all right. He can't go far." Seeing the first dog run away, the second panicked and managed to escape. The dogs were gone and did not come back. Tension ran high in the crew after this mishap with the dogs. The team was worried that they no longer had a way to protect against polar bears. Joël trivialized the problem and seemed to ignore it. Pierre had brought along some security equipment left over from previous expeditions: alarm flares, string, and rubber bands. He devised an anti-bear fence to protect the camp at night.

By deciding to explore a new technique to protect against bears during the preparation phase (i.e., using dogs), the expedition leader placed himself in exploration mode as being responsible for dogs. Once in the field and implementing this technique, the expedition leader entered rationalization mode. They lost the dogs. The effort was a failure. The project leader neither took charge nor responsibility of the situation. Another team member cobbled together (i.e., in adaptation mode) a system using previous skills (i.e., exploitation mode). The effort was a success, but the expedition leader had lost authority.

2.3.2. The ANT Expedition's Critical Situation: Mooring the Ship in the Bay of Whales

For the ANT expedition, the critical situation concerned mooring the vessel in the Bay of Whales. This wintering site was carefully chosen during the preparation phase, because there is no major movement in the Bay of Whales under the ice. This meant that the ship could become iced-in without too much pressure on its hull. The Bay of Whales is 40 meters wide and the ship was 8 meters wide. On each side of the bay are rock cliffs. For the expedition leader and his advisors, the greatest risk was that the ship might smash into the rock cliffs. They were occupied with the question: Once the ship was iced in, how could they prevent it from tilting and smashing into the rocks? A sophisticated system of moorings was devised with the help of several specialists. This strategy was based on the assumption that the hull of the ship would be firmly fixed in the ice.

After sailing for more than three months, the ship moored in the Bay of Whales for a lengthy wintering period of 36 weeks. The moorings were carefully fastened to the rocks, and some were even attached to columns of rock. Time went by, yet weather conditions were such that winter temperatures never came and the boat was still not frozen in the ice. Winter storms raged one after the other with extreme violence. The strength of the winds and the fact that the boat was still floating freely in the water finally caused the moorings to break. The crew got busy repairing them. With the hull of the boat not firmly entrenched in ice, the boat had stretched its moorings to the limit, putting too much pressure on them. The expedition leader was filled with doubt. He considered different scenarios, including evacuation from the bay in case of an extreme problem. One evening, all safety instructions were reviewed with the crew.

Another extremely violent storm hit the bay. The winds were so strong that the moorings gave way one by one. At first, repair teams were sent ashore to repair or strengthen the moorings, but they were breaking faster than they could be repaired. After several hours of struggle, only one mooring remained intact at the front of the ship. “Evacuate the bay,” came the expedition leader’s order from the wheelhouse, thinking he would gather everyone together and pass on the decision, but there was no time. The last mooring at the front of the ship had given way! “Evacuate immediately.” This evacuation plan had been considered for several days due to an interminable series of storms, with winds often exceeding 100 kilometers per hour.

“We’re evacuating? We’re leaving the bay? I don’t agree!” That is how two members of the team who were repairing the moorings questioned the judgment of their leader. However, these crewmembers had only a partial understanding of the rapidly changing situation. They did not have an overview and did not understand the urgency of taking action. These individuals raised doubts about their leader’s decision. However, it was not a mutiny, and they followed orders. It was a matter of knowing who was right. The maneuver was extremely risky. The outcome of events proved that the maneuver was successful. The tension dissipated, and the crew felt renewed confidence in the expedition leader.

During the preparation phase, the plan had been to winter at the Bay of Whales. Specialists were called upon (i.e., exploitation mode). According to plan, the ship moored in the Bay of Whales (rationalization mode). However, an ice pack did not form around the ship because it was too warm. At the same time, the ship was lashed by a storm that broke its moorings. The ship was in danger of smashing into rocks. The expedition leader decided to move the ship out of the bay (i.e., adaptation mode), although he was not the captain (who had left at the very beginning of the wintering season) and was unfamiliar with the area (i.e., exploration mode). The effort was a success, and the leader emerged from the situation with enhanced authority.

2.4. Conclusion

First, all of the modes of action are present to different degrees in both case studies: exploration, exploitation, adaptation, and rationalization. There is a sequence of modes in the action that took place; for the ARC expedition they were exploration, rationalization, adaptation, and exploitation; for ANT, exploitation, rationalization, adaptation, and exploration (see Table 2.1). We can draw an initial conclusion about the relevance of using these two ambidexterity styles to account for how a project operated in a real-life situation.

Second, we cannot reach a conclusion about the relevance of one mode rather than another used by one of the project leaders, nor the sequence or combinations in which such modes were used. In the ARC expedition, it was the choice of the exploration mode during preparation and *in situ* implementation

Table 2.1. Synthesis of the Sequence of Modes of Action in the Two Expeditions

ARC Expedition	ANT Expedition
Exploration	Exploitation
Rationalization	Rationalization
Failure	Failure
No adaptation by the leader	Adaptation by the leader
Adaptation-exploitation	Adaptation-exploration
Success	Success
Leader loses authority	Leader gains authority

in a rationalization mode that led to failure. In the ANT expedition, it was the exploration mode in the implementation phase that saved the boat. However, the choice of an exploitation mode in the preparation phase led the ANT expedition to failure during the implementation phase. Nevertheless, what does stand out quite clearly is that experimentation with different modes and the ability to change modes when necessary is a must. This implies that a project leader must be extremely ambidextrous, whether in the meaning of March or Mintzberg. To understand the use of modes and their combinations, we must refer back each time to the management situation that the project leader is addressing. When the project leader cannot change mode, as is the case with the ARC expedition after losing the dogs, the entire project is at risk. A team member may compensate for the leader's failure to switch modes and suggest an adaptation based on previously acquired competencies that may lead to success. This conclusion is particularly interesting for the entire project management community because it challenges the dogma of rationalization logic during planning, as well as the traditional exploration-exploitation sequence in project plan development, which are currently considered to be the standard.

Third, two questions arise: How is the mode of action chosen according to the situation, and at what point is a change mode needed? It also appears that changing the mode raises questions about a project leader's authority. We are dealing with questions that fall within the scope of organizational learning according to Argyris and Schön (1996) or the learning organization described by Peter Senge (2000), which also raises the question of a perceived shortfall in a project leader in a given situation. This perceived shortfall must be considered as a prerequisite for organizational learning. As showed by Gautier, Lièvre, and Rix-Lièvre (2009), the implementation of an organizational learning process depends on a first stage in which at least one player perceives a shortfall in a given situation that leads to an interpretation in terms of errors; this shortfall may be perceived as a gap between intention and achievement (Argyris, 1993) or between the capacities of the player in question and the situation (Senge, 2000). We see here the exact terms of the dual nature of the ambidexterity required of a project leader as stated in this chapter's summary of the work of March and of Mintzberg. The role of a project leader is becoming increasingly complex, and the competencies related to ambidexterity eventually have to be integrated into project management training programs.

This discussion of ambidexterity, which requires further exploration, highlights a few points: the dual nature of a project leader's ambidexterity, the relevance of studying management situations as soon as possible after they occur to understand the context-based logic underlying the chosen mode of action and its related performance, as well as the need to expand these perspectives by understanding the subjective perceptions of persons involved in a situation.

References

Argyris, C. (1993). *Knowledge for action. A guide to overcoming barriers to organizational change.* San Fransisco: Jossey-Bass.

Argyris C., & Schön, D. (1996). *Organization learning II.* Reading, MA: Addison-Wesley.

Aubry, M., & Lièvre, P. (2010). Ambidexterity as a competence of project leaders: A case study from two polar expeditions. *Project Management Journal, 41*(3).

Benner, M. J., & Tushman, M. L. (2003). Exploitation, exploration, and process management: The productivity dilemma revisited. *Academy of Management. The Academy of Management Review, 28*(2), 238.

Birkinshaw, J., & Gibson, C. B. (2004, June 3). Building an ambidextrous organization. AIM Research WP, EPSCRC, http://www.rhian.net/workingpapers/003jbpaper.pdf

Bredillet, C., Yatim, F., & Ruiz, P. (2010). Project management deployment: The role of cultural factors. *International Journal of Project Management, 28*(2), 183–193.

Brion, S., Favre-Bonté, V., & Mothe, C. (2007, June). *Quelle ambidextrie pour l'innovation continue? Le cas du groupe SEB.* Colloque AIMS, Montréal, Canada.

Chanal V, & Mothe, C. (2005). Comment concilier innovation d'exploitation et innovation d'exploration: Une étude de cas dans le secteur automobile? *Revue Française de gestion, 31,* 173–191.

David, A. (2000). La recherche-intervention, cadre général pour la recherche en management. In A. David et al. (Eds.), *Les nouvelles fondations des sciences de gestion* (pp. 193–214). Paris: Vuibert, Fond National pour L'enseignement de la Gestion D'entreprise.

Farjaudon, A. L., & Soulerot, M. (2008). Les implications du dilemme exploitation/exploration sur le contrôle de gestion: Le cas d'une entreprise de produits de grande consommation. *Cahier du GREFIGE,* Université Paris-Dauphine.

Garel, G. (2003). *Le management de projet.* Paris: Edition La Découverte, Collection Repères.

Garel, G. (2006). Etat des recherches en management de projet. In P. Lièvre et al. (Eds.), *Management de projets, les régles de l'activité à projet.* Paris: Edition Hermés & Lavoisier.

Garel, G., & Rosier, R. (2008, May). *Régime d'innovation et exploration.* Colloque Ambidextrie, IREGE, Annecy, France.

Gautier, A., Lièvre, P., & Rix-Lièvre, G. (2009). Les obstacles en matière d'apprentissage organisationnel au sein de l'organisation de la sécurité civile, une mise en perspective en termes de gestion des ressources humaines. *Revue Politique et Management Public, 6.*

Girin, J. (1990). Analyse empirique des situations de gestion: Éléments de théorie et de méthode. In A. Martinet (Ed.), *Epistémologie et sciences de gestion* (pp. 141–182). Paris: ECONOMICA.

James, W. (1907/2007). *Le pragmatisme.* Paris: Edition Champ Flammarion.

March, J. G. (1991). Exploration and exploitation in organizational learning. *Organization Science, 2*(1), Special issue: Organization learning: Papers in honor James G. March, pp. 71–87.

Midler, C. (1994). *L'auto qui n'existait pas. Management des projets et transformation de l'entreprise.* Paris: Dunod.

Mintzberg, H. (1994). *Grandeur et décadence de la planification stratégique.* Paris: Dunod.

Mom, T. J. M., Van Den Bosch, F. A. J., & Volberda, H. W. (2003). *Managing concurrently the processes of knowledge exploration and exploitation.* Presented at DRUID Summer Conference, Copenhagen, Denmark.

Ney, Favre-Bonte, & Baret. (2008). Vers un modèle de gestion de l'ambidextrie: Innovation d'exploitation interne et coopération d'exploration. *L'innovation organisationnelle: état des lieux, état de l'art.* Symposium conducted at ESC St. Etienne, St. Etienne, France.

Project Management Institute (2013). *A guide to the project management body of knowledge (PMBOK® guide), fifth edition.* Newtown Square, PA: Author.

Rivking, J. W., & Siggelkow, N. (2003). Balancing search and stability: Interdependencies among elements of organizational design. *Management Science,49*(3), 290–311.

Rix, G., & Biache, J. M. (2004). Enregistrement en perspective *subjective située* et entretien en *re-situ subjectif*: Une méthodologie de la constitution de l'expérience. *Intellectica, 38,* 363–396.

Rix-Lièvre, G., & Lièvre, P. (2009). L'observatoire de l'organisant: Mode d'interprétation des matériaux qui en sont issus. *Revue Internationale de Psychosociologie,* Numéro spécial sous la direction de Martine Hlady Rispal, 15, 161–178.

Senge, P. (2000). *La cinquième discipline: Le guide de terrain.* Paris: Editions Générales First.

Tushman, M. L., & O'Reilly, C. A. (1996). Ambidextrous organizations: Managing evolutionary and revolutionary change. *California Management Review, 38*(4), 8–30.

Williams, T. (2005). Assessing and moving on from the dominant project management discourse in the light of project overruns. *IEEE Transactions on Engineering Management, 52*(4), 497–508.

Chapter 3

Mobilization and Sensibility on Polar Expeditions: More than Mere Motivation

Michel Récopé, Pascal Lièvre, and Géraldine Rix-Lièvre

Participants' commitment is essential to the outcome of any project conducted under extreme conditions, such as polar expeditions. The same statement can be applied to the entire field of project management (Baron, 1993; Schmid & Adams, 2008), where outcomes depend more on an individual's actual commitment than personal skills, however crucial they may be. This issue has never been specifically addressed by the management sciences (Garel, Giard, & Midler, 2003).* Some analyses and observations (Récopé, Rix, Fache, & Lièvre, 2006) have shown that a participant's expressed motivation outside the practical context should be distinguished from his or her actual mobilization *in situ*. This calls into question the complex issues surrounding commitment (Kanfer, 1990; Roussel, 2000; Dalmas, 2007). Research conducted in motivational psychology and in the philosophy of experience provides valuable insight into this issue of using notions of mobilization, norms, and sensibility. This research has

* A few articles address the issue of motivation in project management, in particular, Gällsted (2003), Peterson (2007), and Whittom & Roy (2009).

shed light on the behavior of those taking part in polar expeditions, and a case study on two team members with different sensibilities provides empirical support. Our conclusions have led us to reconsider the issue of team recruitment by emphasizing the importance of commitment not only in terms of its intensity but especially in terms of its meaning.

3.1. Contributions from Motivational Psychology and the Philosophy of Experience to the Notion of Commitment

3.1.1. Motivational Psychology

Motivation as a concept can be defined as a "hypothetical construction used to describe the internal and/or external forces that produce a behaviour's activation, direction, intensity, and persistence" (Vallerand & Thill, 1993, p. 17). According to Roussel (2000), this definition provides a fundamental point of reference for management because it is inclusive and avoids the paradigm gaps associated with common definitions. Dalmas, who also favors an inclusive approach, reports a resurgence in interest regarding a participant's own world, along with his or her "self," and believes that improvements in work-related motivation can no longer be addressed without first "identifying the internal levers pushing them [the participants] to act lastingly to commit their energy persistently and in a definite direction" (2007, p. 34). Motivation is expressed through behavioral characteristics associated with commitments made *in situ* (Ryan & Deci, 2000); one would therefore expect evaluation research into motivation to be based on comprehensive qualitative observations made *in situ*. Paradoxically, analyses of studies based on interviews, questionnaires, and scales, all of which share a foundation built on the response of subjects made outside the practical context [e.g., the Sport Motivation Scale (Pelletier, Fortier, Vallerand, Tuson, Brière, & Blais, 1995) or the Passion Scale (Vallerand, Blanchard, Mageau, Koestner, Ratelle, Léonard, & Marsolais, 2003)]. In other words, they rely on judgements, interests, and tastes expressed outside the situation and linked to representations of their interests and tastes. When observation has been used, it was used under experimental conditions or was limited to the amount of time spent on one activity offered from a list of possibilities (Ryan & Deci, 2000); such is the case, for example, with the classic Free Choice Measure (Deci, 1971), when used as an evaluation indicator of intrinsic motivation.

Another paradox is that, although the definitions associated with motivation emphasize the importance of desire, this notion has never been examined in any depth. According to Scherer (2001), the result has been a rudimentary terminology that has led to confusion; in his opinion, goals are linked to any desirable state that mobilizes an organism, without consideration of the source of the motivation or the organism's awareness of it. Lazarus (2001) is more direct: The lack of any effort to identify personal desire is unfortunate. Efforts have been more concerned with distinguishing between intrinsic and extrinsic motivation (Ryan & Deci, 2000), or between mastery goals and performance goals with regard to achievement goals (Dweck, 1986; Nicholls, 1984), despite the accepted notion that intrinsic motivation is rooted in the needs and desires that regulate the direction, intensity, and persistence of an individual's behavior (Deci & Ryan, 1985). Vallerand and Thill (1993) point to internal forces producing these behavioral characteristics. On the other hand, Coquery (1991) stresses the appetitive or aversive values placed on the environmental components targeted by an action. In other words, these authors highlight the adaptive and affective features of motivation but fail to address the basic desires and values involved.

The issue is treated with greater interest by psychologists from the appraisal theory school, who claim that motivational, affective, and cognitive components should no longer be considered separately (Scherer & Sangsue, 2004). The concept of appraisal, clarified in the *cognitive-motivational-relational* theory (Lazarus, 2001), refers to a personal assessment of the environment's significance to well-being—an intuitive and largely unconscious appraisal that is inseparable from the aspects perceived in the here and now. It is the appraisal of the events that guide behavior, not the events themselves. On the one hand, the essential point for Lazarus is that individuals constantly appraise situations according to their personal values (i.e., the existential impact of the goals they set), and they base their actions on this appraisal. The commitment to achieve goals or projects determines the extent of mobilization at the service of what is at stake for the person. The nature and importance of these stakes make up the main criteria used in both appraisal and commitment, and determine, for instance, what constitutes a loss, its significance, and what must be done to prevent it (Lazarus & Smith, 1988). This explains why Lazarus bemoans the lack of research devoted to personal values, goals, and projects that provide the basis of subjective experience and adaptive activity.

This perspective confirms the level of interest demonstrated toward the appetitive and aversive values invested in environmental components (Coquery, 1991), along with the preferential relations the individual establishes with the world: "Certain forms of contact and interactions are preferred to others, certain are sought for and even required (…), others are avoided and apparently harmful" (Nuttin, 1985, p. 15). The logical outcome is that both subject and world do not form two pre-existing and autonomous entities that consequently come together: "the basic unit, from the beginning, is the functional network of the relations themselves (…). Outside this functional unit neither individual nor world exists"

(Ibid., p. 103). Given the significance of relational aspects, any attempt to resolve the observed paradoxes requires a grasp of motivation *in situ,* both within and from the meaningful interaction between subject and world.

Our analysis has made significant progress in considering commitment as the actualization of mobilization and has concluded the following: the individual and the world in relational terms; a permanent appraisal of the event's importance for the individual's well-being as it relates to preferential relations; action is inseparable from the stakes and ensures preferential relations while avoiding harmful ones; cognitive and affective spheres are inseparable; subjective experience is that of a subject experiencing well-being when testing the world.

These advances must be developed further with regard to the actual experience of participants *in situ;* like Carré and Fenouillet (2009), we believe that any conceptual clarification of motivation when connected to action must call upon philosophy. These authors point out that desire and action are central to the great philosophical systems and remind us that human resource managers and leaders seeking the key to individual commitment must contend with these issues.

3.1.2. The Philosophy of Experience

The philosophy of experience that expands Lazarus's position necessarily becomes a philosophy of norms; for Lazarus and Smith (1988), personal norms determine commitment to a goal or to well-being, but these authors fail to develop this notion. We must therefore turn to Canguilhem's philosophy of norms (2003). It is also phenomenological; Lazarus does, in fact, state that his propositions are close to Merleau-Ponty's phenomenology of perception (1945), particularly with regard to the concept of "embodied thought" (Lazarus, 2001, p. 51). The phenomenology of norms* therefore seems better suited to define commitment as mobilization actualized *in situ.*

Norms, Values, and the Living Environment

Canguilhem defines life as a relational activity characterized by appraisal: "Between the living being and its environment, the relationship is established as a debate to which the living brings its own norms with which to appraise the situations" (2003, p. 187). The actual experience of a living being *in situ* is a

* This phenomenology is informed by initial insight for Canguilhem and Merleau-Ponty, two authors whose similarities have recently been reconsidered (Armengaud, 2010; Gérard, 2010), along with their successors.

challenge, in the affective sense of the word, which stems from a relationship to what is normal or abnormal. All norms are an expression of preference: "What is different from the preferable, in a given field, is not what is indifferent but what is repulsive, or rather the repulsed, the detestable. . . . In short, whatever their form, whether implicit or explicit, norms refer to what is real to values, express discriminations of qualities in accordance with the polar opposition of a positive and a negative" (Canguilhem, 2007, pp. 177–178). This experience is the result of the actualization process surrounding one's vital norms, wherein "vital" should be understood as what is experienced as crucial for the individual. This qualifying term applies to preferential relations when their importance is such that one's (good) life is dependent on their satisfaction. This is how all individuals affirm their identity (Macherey, 1998).

According to this perspective, one is never faced with an objective world (e.g., objects, events, circumstances), but each has its own world: the situation, or what has instant meaning here and now and according to our norms and standard. "The proper medium of man is the world of his perceptions, the field of his pragmatic experience in which his actions, oriented and governed by values [...] denote qualified objects, situate them in relation to one another, and all in relation to him" (Canguilhem, 2003, p. 195). Expanding on Merleau-Ponty, Barbaras's phenomenology of life confirms this world of experience: The environment, or everything the organism is sensitive to, is constituted by the organism, "without, of course, this constitution being based on a faculty distinct from the acts by which the living being acts within this medium": the living being "answers the outside world's stimuli in accordance with this organism's proper norms" (1999, p. 143). This is equally the case in neuroscience: Our perception is designed by assigning qualities to external objects, qualities that are dependent on our desires and expectations (Berthoz & Petit, 2006). According to Canguilhem, desires and expectations stem from individuals' own, specific norms.

Thus, the study of experience points to the study of personal sensibility: "The reaction is always determined by the opening of the meaning towards the stimuli and its orientation in relation to them. This orientation depends on the significance of a situation perceived as a whole. Separate stimuli have meaning for human science, but none for the sensibility of a living being. Without recourse to its own vital norms, his action cannot be understood" (Canguilhem, 2003, p. 187). If they do not enter into this perspective, approaches to motivation remains implicitly anchored to a hypothetical *single* environment existing independently of its participants, a ready-made world, a world waiting for individuals and their computations (de Saint Aubert, 2006). Appropriate behavior within this world depends fundamentally on the appropriation of effective techniques and know-how, an appropriation that is more or less correct but nonetheless determines the skills employed by participants. According to this prevailing hypothesis, everyone perceives (more or less clearly) the same external world; subjectivity refers fundamentally to personal tastes and to varying degrees of motivation, which are then applied to this common framework. Personal commitment stems from a modulation, as it is intended to reflect degrees of motivation. Individuals commit either moderately or intensely according to their level of interest in the activity under consideration. This hypothesis entails the existence of a common repertory of executable actions in the world: All participants are (more or less) committed to producing (with varying degrees of success) the same actions. In fact, the same action is likely to be performed under motivation that is either weak or strong, because personal commitment essentially transforms the amount of energy invested consensually. It goes without saying that the phenomenological position disagrees with this hypothesis by supporting the notion that the constitution of the environment is inseparable from the relevant actions, and that their common origins reside within individual norms. These fundamentally represent identity projects and strategies that give rise to a world that matters to an individual, that is, to things one cares about in the world, and that mobilizes one to take action to achieve these projects and strategies.

Norms, Sensibility, and Mobilization

The phenomenological position is clear: The only world is the world of sensibility, because it is an individual's own environment, or the sum total of what sparks an individual's sensibilities according to his or her norms (Barbaras, 1999). The relationship among norms, sensibility, and mobilization must be defined in order to complete our distinction between expressed motivation and actual mobilization. For this, we must first consider desire.

According to Ribot, desiring faculties and sensibilities are equivalent: "sensibility is the faculty of feeling pleasure or pain" (1896, p. 2). Barbaras agrees with Ribot: Desire is at the center of feeling and is very much a *move toward* external objects that are opportunities or conditions of potential satisfaction. According to our analyses (Récopé, 2008), sensibility can be seen as a desiring faculty, but an individual's apprehension of his or her experience *in situ* demands that we define the specific desires involved, that is, the preferential relations established within a given field, which is in this case polar expeditions. The concept of norms is not merely another way to describe specific desires, because it implies a world constituted according to desires. What must be identified is the participant's specific sensibility that promotes his or her sensibility, or what his or her norms have constituted as important objects or events. We must therefore consider the norms that mobilize participants toward what is satisfying or away from what is repulsive. Concepts that associate norms and sensibility account for commitment *in situ,* as well as for character that is inseparable from cognition, affectivity, or locomotion. Considering all three aspects as a whole leads one to disregard motives for actions and to focus on their mobile aspect—hence the term "mobilization." [According to Kant (1989), at an intellectual, moral, and representational level, they can be expressed because they are aroused and guided by reason. However, Kant believed that at a relational, affective, and impulsive level, they reveal sensibility tendencies.] The mobile is a movement, a driving force that Ryan and Deci mention without theorizing: "To be motivated means to be moved to do something. A person who feels no impetus or inspiration to act is thus characterized as unmotivated, whereas someone who is energized or activated toward an end is considered motivated" (2000, p. 54).

3.2. Expressed Motivation and Mobilization during a Polar Expedition

This section presents a case study conducted through observation and participation in a polar expedition whose members were chosen according to their skills. All had expressed a similar motivation for the project. Their varying sensibilities and mobilizations, geared toward the successful outcome of the expedition, surfaced *in situ*.*

3.2.1. Polar Discovery as a Common Motivation

Preparation for the 2004 Greenland Expedition began one year before the actual two-week expedition. The team included four members and one researcher. Three participants were novices at polar ski expeditions. They were, however, experienced in wilderness activities and solo expeditions in high altitude, deserts, and jungles. The expedition leader recruited the team by seeking complementary skills. This case study focuses on two members, both experienced in alpine skiing and climbing but unfamiliar with polar environments. Gérard had training in icy and snowy environments, and was recruited for his ability to travel in these conditions. Dominique, experienced in mountain rescue, was put in charge of safety.

During preparatory meetings, discussions focused on equipment and previous trips. Some shared photographs, whose astonishing scenery and colors delighted the others. Whenever the expedition was discussed, participants

* Names of expeditions as well as individuals have been changed for the purpose of confidentiality.

become motivated by the joy of traveling and discovering the polar environment, not to mention the affordability of the trip.

3.2.2. Observable Evidence of Mobilization

Participatory observation allowed us to examine the way in which these two participants appreciated the same event. Three episodes stood out and shed light on what mobilized each.

A Surprise upon Arrival

In the spring of 2004, the coastal pack ice had already broken up and snow along the sea coast had become scarce; it was no longer possible to ski along the coast as planned. The team had to charter a boat and reconsider how they would access the ice cap; without snow, the pulks could not be loaded or pulled on skis. Two groups were assembled to search for potential access routes. One night, Gérard proposed a route along a stony path, which meant the pulks had to be carried to prevent damage.

The absence of pack ice and snow was experienced in different ways. For Gérard, the event hindered the expedition and wasted time; he prohibited the use of skis, but carrying the equipment made progress slower and more difficult. Other members, however, enjoyed traveling along the broken pack and discovering the glacier's front as it fell into the sea and the groaning sound the ice made as it moved. Dominique took photographs, just as he had done while researching an itinerary. While waiting for sunset, he was astonished to notice that the sun never fully disappeared before rising again.

Gérard and Dominique each had his own specific environment, his own situation and they appraised the absence of pack ice differently. For Gérard, it was annoying and unpleasant; for Dominique, however, the broken pack, the color of the lichens, and the light were magnificent, intriguing, and fascinating. Although each was confronted by the same event, each focused on different aspects and assessed them differently.

Additional information from the case study defines Gérard's mobilization and its orientation and intensity. Upon their return, a debriefing was held at which both Gérard and Dominique were asked to point out two episodes that stood out. Gérard remembered carrying the equipment as being extremely difficult for him: "My morale was not great for those two days. We had to carry a good deal of weight and I was afraid [of hurting myself]." Here, we detect an intensity indicator surrounding his commitment: "I didn't want to hurt myself over something like that." Physical suffering is not conceivable for Gérard in this context. Every other member carried equipment, but nobody else mentioned the episode.

First Installation

Prior to the expedition, tents were to be tested in a snowy environment to guide final selection of a tent. A five-person tent had been borrowed for this test. The most experienced member listed the essential features: a four-season dome tent with two apses.* He owned such a tent, but it had only two places. Gérard promised to obtain one with three places. The two-person tent solution, while not ideal, was approved unanimously. On the first night of the test, around midnight, the team began raising camp. "So," Gérard quipped, "how do you pitch this thing?" He proceeded to unfold the tent while two other members dug a shelter into a snow drift. He called on another member, who was busy pitching the other tent, for advice on putting up his own. That's when they realized they had a three-season tent with two doors but only one apse. Gérard admitted that he had not had time to inspect the borrowed tent.

The tent did not have the required features, but for Gérard this did not matter. A more suitable tent would simply have to be used on the larger expedition.† He helped with the tent, let the others finish raising the camp (i.e., digging out the snow drift, building a snow wall, and making water), and climbed up the crest to locate an itinerary for the next day. He later explained in an interview that back pain had prevented him from digging, so he chose to plan the itinerary instead, hoping to save time on the following day.

For Dominique, the first pitching of the tents was important. The conditions that weekend were not harsh, but they had to prepare for harsher ones where it would be important to avoid contact with snow in order to keep warm. In Dominique's view, Gérard had committed a grave error with his tent. During the interview, Dominique stated that the initial setup had been a series of mistakes: The tent had been pitched in a snow drift in the blowing wind; a missing apse had allowed snow to make its way into the tent; poor collective organization with no distribution of tasks made raising camp difficult. All of these aspects represented weaknesses that had to be rectified before undertaking the larger expedition. The time and energy spent choosing the location, its exposure, and

* Apses are projections of the double roof, forming a sort of lock between the exterior and interior of the tent. This lock can be used to store and protect equipment from bad weather.

† Note that this was a preparatory expedition for a more ambitious project.

digging and protecting the tent—all of these components—Dominique considered necessary to ensure that the camp would be set up properly.

Raising camp held little importance for Gérard: It meant downtime, a lack of progress.

3.2.3. The Financial Investment

The expedition had to be wholly sponsored, and financing was delayed. An advance was needed. The expedition leader paid for the pulks, airline tickets, clothing, and equipment; the experienced member of the team loaned as much equipment as he could. Excess luggage fees and an unexpected boat charter only increased costs. The leader offered to cover these, but financial difficulties arose. Another member, along with Dominique, took care of the additional costs. Dominique also proposed to buy back part of the equipment and to negotiate the sale of photographs, which would fund the expedition. The experienced member offered to pay for his flight and agreed to solicit funds to balance the project's cost. Not once did Gérard offer to advance money, nor did he propose any ideas on how to finance the project.

This represents a significant component about the intensity of Gérard's commitment toward the project. As stated previously, physical hardship mattered to him only when it affected the expedition's progress, and contributing financially to the expedition mattered little to him.

3.2.4. The Issue of Sensibility

Observations made under practical conditions, along with the statements by participants about their experience, explore moments that are significant to sensibility.

At 4 p.m. the break was over and one hour remained to push ahead. The team had to get underway. Gérard urged the team forward and took the lead. He fixed a point on the horizon to maintain direction and keep his bearings. Surrounded by a white and monotonous environment, he oriented himself using color variations in the snow. He identified a lighter point in the distance. This was his landmark, and he never took his eyes off it. He continued to move forward, fixated on this objective. Progress along a snowy, featureless landscape that fades into the distance can become rather dull. From time to time, he would turn his attention to the right or left, but returned quickly to his landmark. He focused on it to guide progress. Seeking distraction, he would look at his skis: They acted as a mirror, confirming his progress.

What had meaning for Gérard was geared toward, and by, progress. The entirety of his sensibility was subordinate to the possibility of moving forward: the color of snow and ice, the landscape, and his skis. After observing him for an hour, at no time did he turn toward his teammates; what was behind him had no relevance.

Dominique broke away from Gérard's tracks and sped up to take advantage of the beautiful light. He photographed the team as it moved forward. He took out his camera, crouched down for a good angle, and snapped a few shots. He then stood up as quickly as possible, put away his camera, put on his overgloves and retrieved his skipoles. He had to protect himself and the camera equipment from the snow, but in his haste to avoid lagging behind, he made contact with the snow. He was preoccupied because the team was gaining distance and he had to catch up.

Dominique was frustrated. He did not have enough time to take the photographs he wanted and had to hurry to avoid being left behind. The pace prevented him from taking full advantage of the beautifully illuminated ice and pristine setting.

3.2.5. Sensibility Differences and Expedition Conduct

The various materials attest to Gérard and Dominique's sensibility and motivation while identifying their specific worlds: Their reactions, actions, and feelings appear to be inseparable from their specific norms, or that which satisfies them. What mattered for Gérard was advancing and progressing as much as possible. Conversely, any time spent not advancing was considered wasted. What mobilized Gérard is known as the "sporting exploit in the wild" type of sensibility (Lièvre, Récopé, & Rix, 2003). Dominique, on the other hand, was mobilized by the "exploration and discovery" type of sensibility (Ibid.): What mattered most was moving forward autonomously while allowing himself to marvel at the extraordinary polar environment.

As part of the same project, their sensibilities diverged despite claiming the same motivation at the outset, to the extent that both became incompatible. Let us consider the consequences for the project. One stems from organizational problems (i.e., regarding the pace of progress and the methods to set up camp). How are we to agree on what constitutes "correct" procedures? Another consequence concerns the difficulty of collectively confronting unexpected events that could endanger the project. The lack of snow and pack ice required adaptive measures that were deemed as a frustrating waste of time by one and as an opportunity for discovery by the other. Further incidents were judged by some as being dangerous and problematic, while others viewed them with indifference. The final consequence involves the manner in which the team was assembled. The expedition leader brought together complementary skills, but the participants' diverging sensibilities and mobilizations, ignored at the outset, became a critical component of great weight.

3.3. Conclusion

This work confirms that an individual's performance in a given project is affected by his or her actual commitment. Theorized as mobilization, and using Lazarus's analyses supported by a phenomenology of norms, this commitment cannot be defined as a matter of intensity but rather as an *orientation toward.* According to this case study, what matters to participants, what provides satisfaction, is what mobilizes them. It is an affective and driving force that pushes a participant toward realizing his or her "good" life. Specifically, commitment is essential to the possibility of mobilizing a particular set of skills. In other words, one's skills, knowledge, and energy cannot be invested into a project unless they correspond to one's norms; neutrality does not count in applying skills, regardless of the conditions involved. Mobilization is inseparable from one's own sensibility, because knowledge, affectivity, and motor skills are all linked; this is in line with Roussel and Dalmas, who call upon an integrated approach to motivation.

The study also makes a distinction between expressed motivation and actual mobilization. The motivation expressed by Gérard fails to reflect his mobilization: What he claimed to value in the project was different from what triggered his sensibility *in situ.* Thus, motivation is a discriminating component when assembling a group for a given project, but it may not be found in the in the group's claims—it surfaces in actions and *in situ.* So, how should one recruit a team to ensure the proper conduct of a given project? One experienced member of polar expeditions suggested a weekend of skiing together for those who are interested, prior to the selection process. According to our definitions, he has an intuitive appreciation for what matters to different participants *in situ* and how they would fit into his organizational framework to ensure a convergence of sensibilities and actual skills.

This study provides insight for management in extreme conditions, and it is also relevant to management situations in a knowledge-based economy (Foray, 2004). The strong likelihood that unexpected and undesirable events may endanger a project (Lièvre & Gauthier, 2009) is linked to extreme management situations, such as polar expeditions. We have demonstrated the critical nature of compatibility regarding personal sensibilities, particularly when confronted by the unexpected. In an effort to collectively adapt procedures and guide decisions when reorienting a project, various team members must be able to share common criteria regarding events. Otherwise, individuals from different sensibilities continue to exist in diverging worlds, which hinders their ability to react in a concerted and timely manner.

References

Armengaud, F. (2010). Georges Canguilhem: Le comportement comme "allure de la vie." In F. Burgat (Ed.), *Penser le comportement animal.* Paris: Maison des sciences de l'homme et Quæ.

Barbaras, R. (1999). *Le désir et la distance.* Paris: Vrin.

Baron, X. (1993). Les enjeux de gestion des salariés travaillant dans des structures par projet. *Revue Gestion 2000, IX*(2), 201–213.

Berthoz, A., & Petit, J.-L. (2006). *Phénoménologie & physiologie de l'action* (*The physiology and phenomenology of action*). Paris: Odile Jacob.

Canguilhem, G. (1952/2003). *La connaissance de la vie* (*Knowledge of life*). Paris: Vrin.

Canguilhem, G. (1966/2007). *Le normal & le pathologique* (*The normal and the pathological*). Paris: PUF.

Carré, P., & Fenouillet, F. (2009). *Traité de psychologie de la motivation*. Paris: Dunod.

Coquery, J.-M. (1991). Motivation. In H. Bloch, R. Chemama, & H. Gallo (Eds.), *Grand dictionnaire de la psychologie* (pp. 480–482). Paris: Larousse.

Dalmas, M. (2007). Les méta modèles de la motivation au travail. *Les notes de recherche du LIRHE 446.* Toulouse: LIRHE-CNRS.

Deci, E. L. (1971). Effects of externally mediated rewards on intrinsic motivation. *Journal of Personality and Social Psychology, 18,* 105–115.

Deci, E. L., & Ryan, R. M. (1985). *Intrinsic motivation and self-determination in human behavior.* New York: Plenum Press.

De Saint Aubert, E. (2006). *Vers une ontologie indirecte.* Paris: Vrin.

Dweck, C. S. (1986). Motivational processes affecting learning. *American Psychologist, 41,* 1040–1048.

Foray, D. (2004). *The economics of knowledge.* Cambridge, MA: MIT Press.

Gällsted, M. (2003). Working condition in projects. *International Journal of Project Management, 21,* 449–455.

Garel, G., Giard, V., & Midler, C. (2003). Management de projet & gestion des ressources humaines. In J. Allouche (Ed.), *L'encyclopédie de la gestion des ressources humaines* (pp. 818–843). Paris: Vuibert.

Gérard, M. (2010). Canguilhem, Erwin Straus & la phénoménologie. *Bulletin d'analyse phénoménologique, 6*(2), 118–145.

Kanfer, R. (1990). Motivation theory and industrial and organizational psychology. In M. D. Dunnette & L. M. Hough (Eds.), *Handbook of industrial and organizational psychology* (vol. 1, pp. 75–170). Palo Alto, CA: Consulting Psychologists Press.

Kant, E. (1788/1989). *Critique de la raison pratique.* Paris: PUF.

Lazarus, R. S. (2001). Relational meaning and discrete emotions. In K. R. Scherer,

A. Schorr, & T. Johnstone (Eds.), *Appraisal processes in emotion* (pp. 37–67). Oxford, UK: Oxford University Press,.

Lazarus, R. S., & Smith, G. A. (1988). Knowledge and appraisal in the cognition-emotion relationship. *Cognition and Emotion, 2*(4), 281–300.

Lièvre, P., & Gautier, A. (2009). Logistique des situations extrêmes: Des expéditions polaires au service secours incendie. *Management & Avenir, 24,* 196–216.

Lièvre, P., Récopé, M., & Rix, G. (2003). Finalités des expéditeurs & principes d'organisation. In P. Lièvre (Ed.), *La logistique des expéditions polaires à ski* (pp. 85–101). Paris: GNGL Productions.

Macherey, P. (1998). *Actualité de Georges Canguilhem* (pp. 71–84). Paris: Institut Synthélabo.

Merleau-Ponty, M. (1945/1995). *Phénoménologie de la perception.* Paris: Gallimard.

Nicholls, J. G. (1984). Achievement motivation conceptions of ability: Subjective experience, task choice, and performance. *Psychological Review, 91,* 328–346.

Nuttin, J. (1985). *Théorie de la motivation humaine.* Paris: PUF.

Pelletier, L. G., Fortier, M. S., Vallerand, R. J., Tuson, K. M., Brière, N. M., & Blais, M. R. (1995). Toward a new measure of intrinsic motivation, extrinsic motivation, and amotivation. *Journal of Sport & Exercise Psychology, 17,* 35–53.

Peterson, T. M. (2007). Motivation: How to increase project team performance. *Project Management Journal, 38*(4), 60–69.

Récopé, M. (2008). Sensibilité. In B. Andrieu & G. Boëtsch (Eds.), *Le dictionnaire du corps* (pp. 300–302). Paris: CNRS Éditions.

Récopé, M., Rix, G., Fache, H., & Lièvre, P. (2006). Sensibilité et mobilisation: Perspectives d'investigation du sens à l'œuvre en situation de pratique. *Revue eJRIEPS, 9,* 51–66.

Ribot, T. (1896). *La psychologie des sentiments.* Paris: Alcan.

Roussel, P. (2000). La motivation au travail. *Notes de recherche du LIRHE, 326.* Toulouse: LIRHE-CNRS.

Ryan, R. M., & Deci, E. L. (2000). Intrinsic and extrinsic motivations. *Contemporary Educational Psychology, 25,* 54–67.

Scherer, K. R. (2001). Appraisal considered as a process of multilevel sequential checking. In K. R. Scherer, A. Schorr, & T. Johnstone (Eds.), *Appraisal processes in emotion: Theory, methods, research* (pp. 92–120). Oxford, UK: Oxford University Press.

Scherer, K. R., & Sangsue, J. (2004). Le système mental en tant que composant de l'émotion. In G. Kirouac (Ed.), *Cognition et émotion* (pp. 11–36). Québec: Les Presses de l'Université de Laval.

Schmid, B., & Adams, J. (2008). Motivation in project management. *Project Management Journal, 39*(2), 60–71.

Vallerand, R. J., Blanchard, C., Mageau, G. A., Koestner, R., Ratelle, C., Léonard, M., & Marsolais, J. (2003). Les passions de l'ame: On obsessive and harmonious passion. *Journal of Personality and Social Psychology, 85*(4), 756–767.

Vallerand, R. J., & Thill, E. (1993). *Introduction à la psychologie de la motivation.* Paris: Vigot.

Whittom, A., & Roy, M.-C. (2009). Considering participant motivation in knowledge management project. *Journal of Knowledge Management Practice, 10*(1), 1–13.

Chapter 4

Mobilizing Social Networks beyond Project Team Boundaries: The Case of Polar Expeditions

Marc Lecoutre and Pascal Lièvre

How can one mobilize relational resources when managing project teams? Many commentators (Akrich, Callon, & Latour, 1988; Midler, 1998, p. 67; Garel, 2003, p. 55) have highlighted how a project leader relies on a network of relationships, and that a project manager without an address book holds a losing hand. Midler termed this the project leader's social competency—a recurring theme in any project management textbook. Paradoxically, this aspect has been conspicuously ignored in most project management literature (Lecoutre & Lièvre, 2010), except by Mead (2001), Huang and Newell (2003), Newell (2004), Julian (2008), and Brookes, Morton, Dainty, and Burns (2006).

We consider polar expeditions projects as ideal natural situations for observing project management in action (Garel & Lièvre, 2010). In doing so, we aim to shed light on the type of resource that social networks represent in the design and implementation of this kind of project. Our aim is to go beyond a straightforward recognition of their role's importance, and to focus deliberately on the question of how such networks are mobilized. Our theoretical framework articulates two research fields. On one side, our study is rooted in the conventional sociological analysis of social networks (Degenne & Forsé, 1994; Lazega, 1998), and specifically the network theory that frames an individual's relational ties as a resource input to his or her actions (Granovetter, 1973, 1995; Burt, 1992), which has gained widespread popularity in management science (Adler & Kwon, 2002; Huault, 2002; Nahapiet & Ghoshal, 1998; Shaw 1999; Swedberg, 2000). On the other side we rely on recent management research into the cooperative mechanisms at work in project teams (Bouty, 2000; Dameron, 2004).

Polar expeditions offer a highly relevant case setting for observing the mechanisms through which social networks are mobilized. First, polar expeditions offer a field that is receptive to the researcher's position, thus enabling the researcher to closely track project goal-completion workflows, and possibly even take on a permanent end-to-end observer-participant position. This is a key argument, because we posit that understanding the sophisticated mechanisms through which social networks are mobilized implies having broad in-field and in-process access to the project action phase, which avoids ex-post rationalizations by the actors involved (Lièvre, Rix-Lièvre, & Lecoutre, 2009). Second, the success of any polar expedition hinges broadly on the contacts established during the project preparation phase, contacts that dictate technical choices and itinerary decisions. We can cite as an example Alain Hubert's Antarctic crossing project (Hubert, 2003, pp. 48–49): While participating in a conference about polar expeditions, he met by chance the physicist Hubert Gallée, who solved his route problem; also, searching high and low in a technical field he did not know at all, he met in a store selling kites a "kite-skiing" enthusiast, with whom he spent a year designing his new sail.

First, we present two polar expeditions that share the same objective but employ highly opposed ways of mobilizing social networks. We then describe how our findings obliged us to deepen the weak ties–strong ties construct, by using research works about community versus complementary modes of cooperation (Dameron, 2004). We opted for a qualitative approach supported by a comparative case study (Yin, 2009). Follow-up was based on a series of formal and informal interviews held with the full project team throughout the project goal completion process (at the project's start, before flying out to the expedition field, then on return back to France). We were also allowed regular email and telephone exchanges with the expedition leaders. Both teams gave us access to all documents produced to present their projects, including the expedition log, photographs, video clips, and more. We also shared a number of informal exchanges with the teams: The two expedition leaders had consulted one author of this chapter as an expert on polar expeditions as well as an expert on the actual expedition site itself.

4.1. Two Case Studies on How Social Networks Are Mobilized for the Organization of Polar Expeditions

To illustrate this topic, we propose two case studies of polar expeditions that shared the same objective (the Spitsbergen crossing), but that differed in the way they mobilized relational network resources.*

4.1.1. Joël's Polar Expedition

Joël is a young civil engineering student. Having spent time living abroad, he caught the travel bug, and naturally volunteered to manage his engineering school's annual overland adventure project as an opportunity to discover the High Arctic. Lacking any relevant experience in polar travel, he began by reading attentively the expedition log left by a team that had led a similar overland challenge project a few years prior. As the expedition leader, Joël recruited the team members himself and determined from the outset who would be going to Spitsbergen and who would be staying in France to track the mission and back up the team. Joël handles this expedition in project management mode.

The first step is sponsorship: With the the help of a team member's father, Joël obtains the backing of a prominent public figure, who enables him to reach a proven polar expeditioner. This specialist's endorsement lends credibility to the novice team, which is embarking for the first time into the polar environment. The second step is finding financial sponsorship: This is well within the competencies of these students versed in marketing techniques. Furthermore, over the years, previous school expeditions had progressively forged a partner network that included former members of the school. The team sets up a website and a promotional information pack during this phase. Joël gathers a budget of around €100,000, a third of which is for expedition gear and equipment.

The third step is groundwork. The team is well aware that their shortage of experience and lack of field knowledge is going to be the sensitive point. This is where they demonstrate their ability to be open and to adapt, by using experts in order "to learn about polar trekking." Here again, drawing on support from their leading public figure, they are given direct access to a team of specialists who take charge of their training for the polar field. Through each successive contact, the project is built up and develops, which leads to a series of on-the-ground training phases involving high-altitude training on weekends with experts with experience in the same expedition crossing. The final two

* Names of expeditions as well as individuals have been changed for the purpose of confidentiality.

months before the expedition's departure has a densely packed timetable that again prompts the team to mobilize resources far outside the core expedition party (e.g., team members' relatives to prepare technical equipment, such as tug-line harnesses for pulkas*). After repeatedly readjusting and fine-tuning the itinerary, the stage-by-stage time table for the journey, the budget, and so on, the decision is made to avoid two dangerous zones—one at the departure point and one at the arrival point—by deploying a mechanical transport solution. The expedition, an entire month on Spitsbergen, takes place and is a resounding success. Once back home, the team is quick to thank everyone who is a partner, sponsor, or supporter of the expedition, while doing media interviews, articles, press conferences, and similar activities.

4.1.2. Luc's Polar Expedition

When it comes to polar expeditions, Luc is a purist for whom the expedition is a philosophy, a way of living. Luc's life revolves around polar travel, and he already has several expeditions under his belt. He is looking to reproduce a style—the early 20th-century long-haul polar voyages. He is also a passionate and accomplished professional mountaineer. Luc is a loner, and when he is planning a trip, his main problem is to find people who are ready and willing to partner with him. Luc plans out his expeditions with a minimum of preparation, confident in his ability to think on his feet and deal with whatever is thrown at him. This makes for tough expeditions—which means tension rides high, and often generates enmity and bad blood with the other team members once the expeditions are over. Two relatives and barely more friends are all that Luc needs to count on as a preparation team to do groundwork. Three months prior to this Spitsbergen expedition, Luc, true to himself, is still without a single team member. Having advertised with various mountaineering clubs, he learns that Eric is interested in this type of project. After a telephone interview, Eric, a novice, finds himself embarked in the project. Luc is setting off for a two-month professional trip in the Southern Hemisphere, but he will email instructions, and sets a date for a first meeting for a weekend's preparatory work, just ten days before the departure.

The budget is stripped down and minimal: one bank loan plus a smattering of personal and family savings. Luc claims his lack of sponsors and refusal to use mechanical vehicles during the expedition as an integral part of his identity. Despite difficulties, Eric and he finally start the expedition from the east coast of Spitsbergen. Just a few days in, they get caught in a violent snowstorm, and their second-hand tent is unable to take the strain. They are forced to backtrack

* A sleigh pulled by a person and used to transport equipment on the snow.

and abandon the expedition. Tension mounts between the two protagonists. Luc finds himself in debt and is facing a seriously compromised financial situation. Eric returns to France. Luc, though, with unwavering moral and financial support from his relatives, elects to stay on site to sharpen his field knowledge and prepare his next attempt at crossing. In Luc's world, all of this is just part of a personal ethic, the trial-and-error learning process, the price to pay for organizing and succeeding in crossing the Spitsbergen in his own way, according to his own values and style.

4.2. Cooperation in Project Management and the Types of Ties Mobilized

At this point, in fairly schematic terms, we can square off two relational approaches to project management: clusters of weak ties purposively networked with the explicit goal of bringing a project to fruition, versus a tight and readily present cluster of strong ties that makes up the major part of resources. This boils down to squaring off "purposive weak ties" against "nonpurposive strong ties." This section discusses the advantages and limitations inherent in these two approaches to mobilizing social networks in project management. This discussion leads to a reassessment of the core principles underpinning weak-tie and strong-tie constructs.

4.2.1. Advantages and Limitations of the Two Network Mobilization Approaches

Joël fully understands the benefits of mobilizing connections well outside his comfort zone in order to acquire project-critical resources, though the dispersal of input sources significantly hampered overall project cohesion. At this point, we find ourselves in the classic strength of a weak-ties scenario as defined by Granovetter (1973, 1982), and extended by Burt's work on structural holes (1992, 2000). It is nevertheless important to underscore the fact that this strategy could not be deployed without the key contribution of strong ties. Joël mobilizes a weak tie: a teammate that he himself had recruited. This weak tie mobilizes a strong tie, his father, who mobilizes in turn another strong tie from his own professional sphere, who then unlocks the possibility of mobilizing radially extended weak ties, which includes a high-profile expeditioner and other mountaineering specialists. This is a textbook case of key weak-tie/strong-tie interplay. Crucially, it is the mobilization of a strong tie that will make it possible for the weak ties already mobilized to broker-in input when called on. So, what prompted Joël to recruit this weak-tie teammate? Joël claims he recruited him based on a shared value set, as well as for his potential contributions to the construction of the project. What makes this tie weak? Let us consider this in Granovetter's framework: The shared-values criterion suggests that this tie should be considered strong. Yet, from the core criterion Granovetter used, that is, the relationship's time frame, it remains a weak tie. In the second case, we characterized the tie connecting the father and his professional connection to the leading public figure as a strong tie, despite the fact that many conventional network theorists (Degenne & Forsé, 1994; Forsé, 1997) would tend to classify it as a weak tie. We nevertheless insist on the strong-tie tag, because these two people have shared several years' experience working together in a professional arena where confidentiality-based relationships are key. So what: strong or weak tie?

In Luc's case, mobilizing strong ties secures him the guarantee of a response when he needs to solicit input, and enables him to hold onto the expedition philosophy he has initiated. Furthermore, in the event Luc has to deal with a major setback, such as the decision to abandon the expedition, his close ties (his "clan") face up to the situation and instantly understand the implications and the issue for Luc. The flip side to this relatively closed network is that it has limited options, as illustrated by the fact that Eric had been recruited as the only candidate; and it accentuates the risk of inadequate support if his clan is solicited for help on technical issues. This core-clan network is by definition unable to face the broad panel of project factor dimensions this type of project requires. There are echoes here of concepts from the sphere of entrepreneurship, where this type of tie offers essentially moral and emotional support. A final point to note is that we identified Eric as a weak tie on the basis that he had never physically met Luc before the Spitsbergen project, which is Granovetter's core criterion for identifying the weakness of a tie. However, their relationship is grounded in a value set that any two mountaineering enthusiasts would share, which shifts their tie closer to what Granovetter would term a strong one. So what makes a tie strong? We have now to look back to Granovetter's work and to consider his pivotal distinction between strong and weak ties, before reassessing the interplay of strong ties and weak ties found in both cases.

4.2.2. Cooperation: Strong Tie or Weak Tie?

The problem hampering our analysis lies in the distinction between strong and weak ties. Let us return to the criteria that Granovetter used to build this partition. We have now to drill down into the strong tie/weak tie constructs and to examine in detail what exactly makes a tie strong or weak. Granovetter (1973, p. 1361) asserts that the strength of a tie between two individuals is reflected in four combined factors: the duration of the link, the emotional intensity, the level of intimacy, and finally, the nature of the services exchanged. He accepts a degree of interfactor independence, but the most telling point is that the hierarchy he established is non-neutral. We explore this point further, before considering how these factors interplay.

The first factor differentiating weak ties from strong ties is the duration of a relationship. This relational time frame can be conceptualized in three different ways:

1. How far back the relationship stretches: “I’ve known him for 30 years.”
2. The amount of real time spent together: “We played alongside each other on the same basketball team for three-and-a-half years.”
3. How often they met: “I used to see him three times a month.”

These different condition sets can be combined. This makes it possible to use these different variables to square off strong ties against weak ties without necessarily proposing a value of an indicator that clearly distinguishes between weak-type and strong-type ties. Time frame is a criterion that is easy to use as a metric, and Granovetter draws on it most heavily (1973, 1982, 1995), often to the exclusion of the other three.

We can combine the second and third factors, putting together the emotional intensity experienced directly by the actors, which should probably be further qualified as *positive* emotional intensity, with the intimacy factor, which is based on the degree of mutual interpersonal trust. Granovetter uses these criteria to qualify the proxemic degree of closeness perceived between individuals. This proxemic degree of closeness reflects how an individual subjectively assesses other people. It is when one person identifies another as very close that they develop a relationship. Intimacy and positive emotional intensity may express a mechanism of interpersonal identification through common values. However, when we consider a tie to be strong because it exhibits intense interpersonal identification, we are also considering one of the two mechanisms founding the cooperation according to Dameron (2004), which in her research about cooperation within project teams is called communautarian cooperation. The nature of a tie is defined by how the actors mutually identify with each other. This point is a clear demarcation line between these two factors, the emotional intensity and the degree of mutual trust, and Granovetter’s fourth and last factor, that is, the type of services exchanged, that we analyze now.

So Granovetter’s fourth and final factor covers the type of services exchanged. These services can be viewed as a global indicator of reciprocal informational and service exchanges. They can also clearly be viewed as the second mechanism underpinning cooperation, which is based on the mutual exchange of resources among people acting in a perspective of calculated self-interest, which Dameron (2004) calls complementary cooperation.

These two sources of cooperation must, analytically at least, be clearly segregated. As shown by Dameron, these two mechanisms refer to big parts of sociology, running from Durkheim to Crozier and back to Weber. In our aim to categorize the intensity of social ties, we feel it is legitimate to distinguish the type of cooperative framework involved, because each cooperation mechanism—communautarian or complementary—should match a type of relationship. Furthermore, our analysis has a pragmatic perspective: An actor's ability to draw on a large number of social ties is relevant only if these ties do actually generate cooperation with him. Why develop network ties if they don't respond when called on? We are clearly in Granovetter's perspective, where a network is considered only as a resource for action.

Consider now the case of a very weak tie according to the three criteria: time frame, degree of shared-identity closeness, and level of information or service exchange. First, let us look at the time-frame variable: I may have spent three hours yesterday with a given person for the first time in my life, and nevertheless begin today to exchange much information with that person just because we realized after a few minutes that we shared a degree of mutual connivance. An exceptionally short relational time frame is not in itself a barrier to cooperation. Second, if we take the shared-identity criterion, if there is only a very weak degree of closeness, there will be no initiation of community cooperation; and, if we take the third criterion, if there is only a very low potential for fruitful exchange, there will be nothing for initiating complementary cooperation either. In plain language, this means that the cooperation perspectives, that is, the beginning of a relationship, for this type of tie are virtually nonexistent. So what kind of weak tie is this? The weak tie according to Granovetter? We believe differently. We posit that the Granovetter weak tie may indeed be weak according to the time-frame criterion, but this tie is in fact potentially strong according to one of the two other criteria capable of initiating actual cooperation. Focusing on this type of weak tie, as it is the one that responds when mobilized, we have recast it as the "*potentially cooperative weak tie*" (Lièvre & Lecoutre, 2006). Its potential for cooperation to take place may be activated through either perceived co-identity closeness or through the possibility of exchanging information or services. This distinction, from an instrumental standpoint, is fundamental.

Dameron (2004) previously showed the benefits of differentiating these two cooperative frameworks within the project team's cooperative mechanisms. More important, she demonstrated the possibilities for switching between these two kinds of cooperation within the same interpersonal tie as a source for securing actual and sustainable cooperative practices. It is easy to imagine that these two mechanisms of cooperation are linked in the same social relationship, that the weakness of one may be counterbalanced by the strength of the other, and that this articulation may evolve over time. Either way, one takeaway is that mobilizable weak ties are *potentially cooperative weak ties,* regardless of the source for cooperation. And these potentially cooperative weak ties are also, by definition, potentially *strong* ties.

4.3. Three Illustrations of Weak Ties and Cooperative Potential

Three examples of weak ties and cooperative potential borrowed from the stories of Luc and Joël illustrate this concept of a *potentially cooperative weak tie*. The aim is to show where triggers of cooperation exist and do not exist in the process of mobilizing social networks in a project. The main problem with a weak tie is that there are no guarantees it will respond when called on, which is what makes it so important for a weak tie to be potentially cooperative.

4.3.1. A Weak Tie without Cooperative Potential

Joël has to get in touch with a company that, three years ago, had heavily subsidized a graduate-student expedition to the Andes. However, the then-CEO has since moved on, and Mr. Paul is now head of the company. Joël feels confident before his first meeting with the new boss: His engineering school has long enjoyed an excellent reputation at the company. Despite this, the first contact is somewhat cold. As he delivered his project, Joël had the feeling that he and Mr. Paul are "worlds apart." After restating his high esteem for Joël's school, Mr. Paul cuts straight to the point, declaring in no uncertain terms that he cannot see how the next generation's engineers could reap any benefits from a course project as "whacky" as a polar expedition. Mr. Paul also advises Joël that he should stop thinking about exotic holidays and knuckle down to do serious work. He further states that he also struggled to see what benefit his company could gain from sponsoring an event like this, which he feels is totally at odds with a sound corporate image. This weak tie mobilized by Joël inititates no identity-based cooperation because Mr. Paul and he are worlds apart. Nor does it unlock any utilitarian-based cooperation, because Mr. Paul cannot see how his company could expect to reap any benefits from endorsing Joël's project.

4.3.2. A Weak Tie Potentially Cooperative from a Utilitarian (or Complementary) Stance

When Hervé meets Joël at *Nuits Polaires*, the polar expeditioning travel fair in Paris, he instantly sees the benefits of sponsoring an expedition that enjoys heavy media exposure and is backed by substantial endorsements, because Hervé has just set up his own polar travel agency. Hervé, then, is most attentive when Joël goes on stage to present his expedition project. The project start-up phase has just begun. Joël is hunting for sponsors to provide financing and equipment. He has already booked airtime on a national radio station, a TV channel, and space in three national and local newspapers. Hervé follows through with a proposal to help by loaning them the expedition gear and equipment with the agency logo. Joël happily accepts. In this example, the nature of the weak tie beginning to connect Joël and Hervé is clearly and intentionally based on complementary cooperation.

4.3.3. A Weak Tie Potentially Cooperative from a Shared-Identity (or Communautarian) Stance

Luc recruits Eric based on shared-identity cooperation. Luc and Eric were complete strangers at the outset. Eric becomes aware of Luc's project through a small ad card left at a local mountaineering club. The message is short and to the-point: "*Wanted: teammate for crossing the Spitsbergen.*" Luc added his phone number underneath the message. For years, Eric, an experienced mountaineer, has nurtured the dream of trekking through the High Arctic. Furthermore, the timing also works, because a ski trek with friends that is planned for April in the Southern Alps has just been canceled, because Eric's friends could not get time off work. He finds himself available. Everything points to a dream opportunity. He makes the call that same evening. After just a few sentences on the phone, Luc and Eric feel they are on the same wavelength: They speak the same language, the one of mountaineers.

After thirty minutes on the phone, it is as if Luc and Eric are already old friends. The stage is set: Luc and Eric will set out together for Spitsbergen. Eric is even slightly taken aback by the level of trust that Luc conferred on him from the outset, when he entrusted him with the expedition groundwork for the next two months because Luc has to be away on business abroad. Nevertheless, Eric finds Luc to be "a really cool guy!" Not only is Eric "on board," he is also recruited straightaway as the lead project organizer. Because they were both able to identify with the same set of values, Luc and Eric immediately launched into cooperation to build a project together.

4.4. Conclusions: Perspectives on Entry into Cooperation in Project Management

We propose two points to conclude. First, the interplay between weak ties and strong ties can now be explored on a deeper level based on the more precise definitions of strong and weak interpersonal ties we proposed above. On one hand, Joël's project was essentially built on weak ties, but it exploited decisive input from project team members' strong ties (i.e., mobilizing a leading political figure). On the other hand, Luc's project was anchored by many strong ties except in the crucial teammate recruitment process, where Luc was left with no option but to turn to a weak tie. What is this interplay becoming? Distinguishing strong ties from weak ties through the time-frame factor is not enough to account for it. What is significant in mobilizing the chain of relationships is in fact the cooperative nature of the ties connecting actors. Could there be relational chains in which discontinuous forms of cooperation would emerge? This is what the example of Joël's project demonstrates. A strong tie with close shared identity enabled Joël to mobilize a tie that is weak from a complementary cooperation stance. Here we find again Dameron's theory, which states that there is a challenge in stimulating articulation between the two forms of cooperation to leverage the cooperation activity itself. In Luc's case, mobilizing a weak tie that is potentially cooperative from the community stance worked out well: He found a teammate with whom he was effectively able to share part of his expedition, though the tie is stretched to the breaking point by the end of the project. Furthermore, following Bouty (2000), the ideal situation would be to mobilize ties that are weak from a time-frame perspective but that offer strong potential for shared identity–based cooperation, so an all-round well-balanced cooperation about exchange of information and services could develop. Reciprocally, service exchanges can provide the platform for long-term sustainable cooperation, what Kreiner and Schultz (1993) have termed the barter economy. Taken together, these findings strengthen the idea that the underlying mechanisms for mobilizing social networks need to be further examined by scrutinizing weak tie/strong tie constructs and the conditions governing their interplay.

Our second and final point addresses the thorny issue of intentionality. Can a person's relational network be maneuvered explicitly in an accumulative strategy, with an essentially instrumental purpose, or should we consider the benefits garnered as relatively contingent subproducts of social connections (Steiner, 1999)? Instrumentalizing a relationship too obviously necessarily has an impact on the relationship itself. The interaction may simply be seen as a straightforward, one-off exchange. An individual who adopts an inflexibly utilitarian stance risks having his actions quickly seen as manipulative, which will sooner or later kill off the relationship he had sought to develop. There is no denying that one engages readily in actions with others on the basis of the feeling that you share common values. Paradoxically, it is only through this experience that people can, in turn, develop their perception of one another and determine that values are actually shared. In our case studies, the positive spin-off from relational ties stemmed not only from intentional investments but also from interpersonal understanding that extended beyond the bounds of strictly instrumental objectives that prompted the actors to initiate contact. Our analysis sheds new light on the intentionality issue: As shown by Dameron (2004), it is possible to play on the crossover between the two forms of cooperation. Again, the shared-identity dimension of cooperation highlights limitations we evoked: Stretching the strategic option to the limit kills off the network because suspicion arises, whereas acting without any purposive strategy foregoes a powerful resource for action.

References

Adler, P. S., & Kwon, S.-W. (2002). Social capital: Prospects for a new concept. *Academy of Management Review, 27*(1), 17–40.

Akrich, M., Callon, M., & Latour, B. (1988). A quoi tient le succès des innovations? [What is the succes of innovations?] *Gérer et comprendre, annales des mines, 11,* 4–17 & *12,* 14–29.

Berry, M. (1996). Savoirs théoriques et gestion [Theoretical knowledges and management]. In J.-M. Barbier (Ed.), *Savoirs théorique et savoirs d'action [Theoretical knowledges and Action knowledges*] (pp. 43–56). Paris: Presses Universitaires de France.

Bouty, I. (2000). Interpersonal and interaction influences on informal resources exchanges between R&D researchers across organizational boundaries. *Academy of Management Journal, 43*(1), 50–65.

Brookes, N. J., Morton, S. C., Dainty, A. R. J., & Burns, N. D. (2006). Social processes, patterns and practices and project knowledge management: A theoretical framework and an empirical investigation. *International Journal of Project Management, 24*(6), 474–482.

Burt, R. S. (1992). *Structural holes: The social structure of competition.* Cambridge, MA, & London, UK: Harvard University Press.

Burt, R. S. (2000). The network entrepreneur. In R. Swedberg (Ed.), *Entrepreneurship—The social science view* (pp. 281–230). Oxford, UK: Oxford University Press.

Coleman, J. (1990). *Foundations of social theory* (pp. 300–321). Cambridge, MA: The Belknap Press of Harvard University Press.

Dameron, S. (2004). Opportunisme ou besoin d'appartenance? La dualité coopérative dans le cas d'équipes projet [Opportunism or need to belong? The cooperative duality in the case of project teams]. *Management, 7*(3), 137–160.

Degenne, A., & Forsé, M. (1994). *Les réseaux sociaux* [*Social networks*]. Paris: Armand Colin.

Forsé, M. (1997). Capital social et emploi [Social capital and employment]. *L'Année sociologique, 47*(1), 143–181.

Garel, G. (2003). *Le management de projet* [*Project management*]. Paris: La Découverte, Collection Repères.

Garel, G., & Lièvre, P. (2010). Polar expedition project and project management. *Project Management Journal, 41*(3), Special issue: Project management in extreme environnements, 21–31.

Granovetter, M. S. (1973). The strength of weak ties. *American Journal of Sociology, 78,* 1360–1380.

Granovetter, M. S. (1982). The strength of weak ties: A network theory revisited. In P. V. Marsden & N. Lin (Eds.), *Social structure and network analysis* (pp. 105–130). Beverly Hills, CA: Sage.

Granovetter, M. S. (1995). *Getting a job: A study of contacts and careers,* 2nd revised ed. Chicago: University of Chicago Press (1st ed., 1974).

Huang, J. C., & Newell, S. (2003). Knowledge integration process and dynamics within the context of cross-functional projects. *International Journal of Project Management, 21*(3), 167–176.

Huault, I. (Ed.). (2002). *La construction sociale de l'entreprise. Autour des travaux de Mark Granovetter* [*The social construction of firms. Around the work of Mark Granovetter*]. Caen, France: Editions EMS—Management et Société.

Hubert, A. (2003). Eléments de logistique en matière d'expédition polaire à ski [Elements of logistics for polar ski expedition]. In P. Lièvre (Ed.), *La logistique des expéditions polaires à ski* [*Logistics of polar ski expeditions*] (pp. 46–51). Paris: GNGL.

Julian, J. (2008). How project management office leaders facilitate cross-project learning and continuous improvement. *Project Management Journal, 39*(3), 43–58.

Kreiner, K., & Schultz, M. (1993). Informal collaboration in R&D: The formation of networks across organization. *Organization Studies, 14*(2), 189–209.

Lazega, E. (1998). *Réseaux sociaux et structures relationnelles* [*Social networks and relational structures*]. Paris: PUF.

Lazega, E. (2000). Rule enforcement among peers: A lateral control regime. *Organisation Studies, 21,* 193–214.

Lecoutre, M., & Lièvre, P. (2010). Mobilizing social networks beyond project-team frontiers: The case of polar expeditions. *Project Management Journal, 41*(3), Special issue: Project management in extreme environnements, 57–68.

Lièvre, P. (2004). La logistique en milieux extrêmes, le cas des expéditions polaires [Logistics in extreme environments, the case of polar expeditions]. In P. Lièvre & N. Tchernev (Eds.), *La logistique entre management et optimisation* [*Logistics between management and optimization*] (pp. 149–156). Londres, France: Edition Hermès Lavoisier.

Lièvre, P., & Lecoutre, M. (2006). Le processus de mobilisation des réseaux sociaux dans une démarche de projet: La notion de lien faible potentiellement coopératif, une application au cas des expéditions polaires [The process of mobilizing social networks in a project approach: The concept of potentially cooperative weak tie, an application to the case of polar expeditions]. *Revue sciences de gestion, 52,* 83–106.

Lièvre, P., Rix-Lièvre, G., & Lecoutre, M. (2009). *Une proposition méthodologique d'investigation du réseau en acte* [*A methodological proposal to investigate social networks in action*], communication à la 3ème journée Management et Réseaux Sociaux, AGRH-AIMS, IREGE, Annecy, 7 novembre.

Mead, S. (2001). Using social network analysis to visualize project teams. *Project Management Journal, 32*(4), 32–38.

Midler, C. (1998 [1993]). *L'auto qui n'existait pas. Management des projets et transformation de l'entreprise* [*The car that did not exist. Project management and transformation of the firm*]. Paris: Dunod, coll. Stratégies et management.

Nahapiet, J., & Ghoshal, S. (1998). Social capital, intellectual capital, and the organizational advantage. *Academy of Management Review, 23*(2), 242–266.

Newell, S. (2004). Enhancing cross-project learning. *Engineering Management Journal, 16*(1), 12–20.

Shaw, E. (1999). Networks and their relevance to the entrepreneurial/marketing interface: A review of the evidence. *Journal of Research in Marketing and Entrepreneurship, 1*(1), 24–40.

Steiner, P. (1999). *La sociologie économique* [*Economic sociology*]. Paris: La Découverte, coll. Repères no. 274.

Swedberg, R. (Ed.) (2000). *Entrepreneurship. The social science view.* Oxford, UK: Oxford Management Readers, Oxford University Press.

Yin, R. K. (2009). *Case Study Research: Design and Methods* (4th ed.). Beverly Hills, CA: Sage.

Chapter 5

A Methodology for Investigating the "Actual" Course of a Project: The Case of a Polar Expedition

Géraldine Rix-Lièvre and Pascal Lièvre

The purpose of this chapter is to present a methodology for investigating the "actual" course of a project, in this case a polar expedition. We are therefore working within the "project as practice" framework (Blomquist, Hällgren, Nilsson, & Söderholm, 2010). This system investigates the practices of actors in terms of Bourdieu (1977), that is, practices expressed strictly *in situ*, and of the Chicago School (i.e., the work of Mead, Blumer, and Strauss), which articulates individual and collective concerns from an "interactionist" standpoint. By attempting to point out the collective action of organizing in its full actualization, this observatory follows roughly that used by Weick (2003). We investigate organizing and study the conditions through which it occurs by attempting to resolve the problems associated with the study of the activity itself, as well as considering the individual and collective dimensions of the organizational dynamic. In the words of Karl Weick (2003), we must try to understand "how organizational life unfolds," specifically *in situ*. How do the actors, individually and collectively, construct meaning for their actions, and what organizational dynamics are used? While classic methodology focuses primarily on "ways of saying" (Hlady Rispal, 2002), the qualitative methodology proposed here belongs to a new approach that focuses on "ways of doing" (Rix-Lièvre & Lièvre, 2009).

In our attempt to address the difficult question of how to pass from an individual focus to a collective one, and vice versa, we propose an observatory consisting of two complementary investigation tools: the Multimedia Logbook (MLB), which focuses on the collective, and the Situated Practices Objectifying System (SPOS), which focuses on the individual. Since each tool requires the personal and specific involvement of a single researcher, the overall system therefore depends on the simultaneous involvement of two researchers.

The system is based on methodological considerations that date to 2000 and were part of a logistical research program for extreme situations, specifically polar expeditions. Polar expeditions are considered project activities (Garel, 2003) with exemplary characteristics for research because:

1. The associated context provides for more readable phenomena because the logic of the actors is pushed to the limit.
2. They allow for participant observation that is as close as possible to the situations experienced by the actors.

First, we list the obstacles that arise when investigating actual organizing as it occurs. Second, we present the observatory by defining each of its tools, and discussing how they work together and complement one another.

5.1. The Obstacles to Overcome when Investigating Organizing as It Is Occurring

Before we present the methodology that was designed to investigate organizational actualization in polar ski expeditions, we must first reconsider the inherent limits that using only observation or interviews imposes on the understanding of practices. This section first calls attention to the limits of observation when studying practices. It then demonstrates the limits of discourse when seeking to understand activity.

5.1.1. The Limits of Observation

In light of the paradigmatic shift addressed by Dosse (1995) in the social sciences and by Rouleau (2007) in organizational theories, it appears that actions can no longer be considered beyond their meaning. Behavior is considered as "an object of study with two faces—one public and behaviorally observable, the other private and unobservable" (Vermersch, 2004, p. 36). An actor is thus no longer merely an object of study, but rather the subject itself. He or she should by no means be considered "a cultural idiot" (Garfinkel, 1967), but rather an intelligent, rational, and occasionally sensitive subject whose subjectivity is worthy of enquiry.

Since Malinowski (1963), anthropology has promoted participant observation, that is, a researcher's immersion into a specific group to understand the way in which the actors live and how they see the world. By separating from his or her daily reality and spending time with a given population while participating in its activities, a researcher is able to construct a way in which others understand their physical and human environment. Participant observation therefore seeks to understand both the observable and the subjective sides of behavior; Lévi-Strauss, working along these lines, wrote that "to properly understand a social reality, it must be completely understood, that is, from the outside, as a thing, but as a thing that fully integrates the subjective knowledge (conscious or unconscious) we would gain if, necessarily human, we were to experience it as natives, rather than observe it as ethnographers" (1950, p. 28). Despite the ethnocentric risks involved, the issues of experience and meaning therefore appear to have been addressed and resolved by using participant observation.*

Despite the value of participant observation when seeking to understand practices as they occur, we intend to go further, as in ethnomethodology. Actors' ability to describe and interpret their own practices spontaneously must be considered. The description of a situation and of the activity should not be "monopolized by the scientific observer [. . . but must be] realized from the point of view of the actor's internal dynamics" (Theureau, 2000, pp. 182–183).

It is therefore of interest to question actors about their actions and activities from their own point of view, as well as about the way in which they consider their practices.

5.1.2. From the Limits of Spontaneous Discourse when Studying Activities to a New Context of Expression

Even if we were to consider the actors' point of view by asking actors to discuss their practices, their statements would do little to clarify these activities in their

* It should be noted that the term "participant observation" covers a wide range of research methods (Ghasarian, 2004; Soulé, 2007). It appears to be necessary, at the very least, for researchers to include the basis for these methods in their own research, as we will do later in this chapter.

totality. Depending on the context, actors develop different ways to define their actions (e.g., explaining, justifying, evaluating, and describing). These somewhat spontaneous verbalizations vary in nature and exhibit specific links to each action. In our effort to understand the practices involved in polar expeditions, we must distinguish among the different types of verbalization about these activities and provide different contexts of expression that allow us to understand, on the one hand, the dynamics of the collective action and, on the other hand, the basis of an action from the actor's point of view.

Spontaneous remarks tend to present an actor's actions as "a normal example of an action normatively organized" (Quéré, 1993, p. 69). Here, an activity is made rational or acceptable when presented to others in a given social context. This way of stating one's actions is not calculated but represents the natural semantics of an action. In the more formal context of a sociological interview, actors are led "to talk about themselves and to select what they consider to be meaningful traits from their past" (Lahire, 2002, p. 391). The resulting discourse stems from a "verbal construction of oneself by oneself [which] is the product of narrative work based on the observation of oneself by oneself and by others" (Ibid., p. 392). These verbalizations therefore resemble a plot (Ricœur, 1983): an ordering that presupposes logic or causality in the succession of events. They reveal the coherence conferred *a posteriori* upon an actor's actions and what the actor wants others to see. As a result, they allow us to specifically grasp individuals' constructed identities (Dubar, 1991), to understand the knowledge they use when justifying their practices, and to study the goals they set for themselves. They do not, however, provide any understanding of the action's rationality *in situ,* that is, the logic that governs the action as it actually occurs.

In fact, this ordering does not involve rationality-in-action. "Action is a form of knowledge in its own right, [. . .] it exists, it functions, it has goals and it reaches them, without necessarily being conceptualized" (Vermersch, 1996, p. 72). The logic that governs action represents knowledge-in-action, that is, an embodied meaning, a working knowledge that can only reveal itself in an action itself, from the action and during the action. Therefore, its content can only be qualified as embodied, antepredicative, or implicit insofar as it has not yet been consciously conceived. An action's rationality and that of discourse are therefore different in nature. *Acting* is characterized by its efficiency along with its practical and actual relevance in a given moment; it is a creation, an implementation. *Saying* is a demonstration and a presentation; it is characterized by its coherence and its overall scope. There is a true epistemological rupture at work between acting and saying. Faced with the impossibility of reducing practice to discourse, a researcher must contend with a dilemma. Either he or she ceases to treat an action as the subject, insofar as science can only progress through discourse, or he or she accepts an action's specific nature along with the impossibility of explaining it only through verbalizations, while attempting to understand and explain it by focusing on the relationship between the actor's verbalizations and actions. A researcher must therefore examine the production conditions surrounding verbalizations that document acting, that is, the logic of the action, the working knowledge-in-action, or the way in which an actor constructs and experiences his or her situation.

On the basis of these two statements, our observatory attempts to combine both levels of understanding—observation material and discourse—in an effort to study the way in which actors experience and behave during an expedition.

5.2. The Observatory

As we mentioned in the introduction, our objective is to investigate organizing in its full actualization by attempting to resolve the problems associated with studying activity and by taking the individual and collective dimensions of the organizational dynamics into consideration. Accordingly, implementing this observatory requires two researchers, each with his or her own specific objectives, roles, and investigation tools. Both take part in the same expedition from conception to preparation, from undertaking to return. Both must be integrated into the team, but each has to construct, in the anthropological sense of the word, a different fieldwork. The following presents the work of both researchers in succession, along with their roles and investigation tools, using investigation examples obtained during a polar expedition in Labrador. Finally, we show how both systems and constructed materials complement one another in order to understand the actual course of an expedition.

5.2.1. Considering a Collectively Shared Reality: The Multimedia Logbook

As an experienced member of polar ski expeditions, the first researcher's role was that of an actor. Like the other team members, he took charge of certain responsibilities (e.g., the tent, the stove, bear protection, or forward progress). He acted, intervened, and assumed part of the leadership. Once again borrowing from anthropology, this approach to understanding practices *in situ* may be qualified as "observing participation," as in Junker (Peretz, 2004), which points out that a researcher's observations are subject to his or her activities as a participant. However, the actors were informed of the study's methods and purpose. They took part willingly and received reports. David (2000) discusses direct participation when categorizing this type of participant observation. Direct participation does, however, imply that the actors expected the researcher to provide a level of physical and moral commitment equal to their own. This level of commitment, along with responsibility for constructing materials, cannot be assumed of any researcher, especially during an autonomous and risky polar ski expedition, unless he or she possesses a certain expertise in the activity itself. For this reason, the researcher who assumed this role during the Labrador expedition had more than 15 years of experience in polar expeditions. He had acquired skills for progression, raising camp, and bear protection, and he also understood what information could be gained, knew experts who could help implement the project, and what technical equipment was available on the market. Thus, regardless of the expedition's schedule or the phase of the project, this researcher was fully able to confront the various situations involved.

However, the notion of observing participation leads to another consequence. Because a participating observer must give a significant level of commitment, such an observer may have to intervene. In fact, such a researcher does not attempt illusory neutrality or passive observation of a so-called natural situation but rather assumes a willingness to intervene in certain cases (Berry, 2000; Plane, 2000). Therefore, not only does this type of researcher take part in all of the group's activities and choices, he or she also provides the group with development, monitoring, and adjustment tools, as well as intervenes at certain times to ensure progress. For example, after the Labrador expedition lost the dogs used to warn and protect against bears, the researcher proposed a makeshift alternative involving a fence around the camp. This observing participation allowed the researcher to develop a certain type of interiority within the group and its activities. This interiority is essential when approaching collective decision-making processes, strategic thinking, and project management. Acting as an authority on anti-bear protection during the preparatory phase of the expedition, the researcher's activity shed light on the decision-making process, along with its foundations and potential risks. In that moment, the researcher found himself in a legitimate position to question the need to include a firearm in each tent, that is, two firearms in total. This questioning, along with the reaction of the expedition leader, who, fearing danger, swept this option aside, provided an opportunity to define the way in which the expedition leader constructed his authority.

With this ability to approach internally the operations of a group pursuing a project, this researcher, like any such researcher who takes on the role of an actor, had to develop, above all, investigation tools that allowed him to construct a record of the group's activities despite his involvement in it. Throughout every phase of the expedition, from the initial idea to the forming of the team, from the search for sponsors to the choice of equipment and preparations, from departure to the return to France, the researcher had to be able to assemble and preserve the various elements that recount the events that occurred within the group.

To conduct this longitudinal study, a Multimedia Logbook (MLB) was developed. It used a number of different formats and assumed many forms depending on the phase involved. During the team assembly phase, the MLB included emails that were exchanged among the various participants, along with notes and recordings made by the researcher during meetings. It also included video to help the researcher create an unbiased record of his daily progress during ski treks and, most important, the group's organization and specific events. Together these various materials ultimately created a point of view that corresponded to the researcher's during the project.

Using this logbook and self-reflection regarding practices, as described by Vermersch (1999), the researcher produced a film and a written account of the expedition. Materials contained in the logbook were based on the researcher's experience as an actor within the group. Using these materials to construct scientific knowledge, from a constructivist role, the researcher had to reflect *ex post* in order to describe his experience and the way his observations were developed (Le Moigne, 1995). This reflection can be linked to the researcher's indispensable reflexivity, as described by various authors such as Geertz in anthropology and, more broadly, Strauss in the social sciences. The resulting reflections allowed the researcher to consider his investigation method and to create distance from his own experience. Consequently, he gained the ability to create an account disconnected from his own subjectivity as an actor who was seeking acceptance from the other members of the team. As a result, the account in the MLB had consistency between various components or events contained in its story—a story that recounted the various phases of the expedition by creating logical links between them (Ricoeur, 1983).

On one hand, this account, which tells of a reality that was shared across the entire scope of the expedition, forms the corpus through which the researcher may question organizational methods by focusing on specific aspects, such as recruitment, or monitoring the leader's role throughout the project. On the other hand, the account is a source of information for all practitioners. By accessing the actual course of collective organization *in situ,* practitioners may extract lessons that can be applied to their own project.

5.2.2. Entering Each Actor's World of Practice: The Situated Practices Objectifying System

The first researcher's investigations were directed at the collective side of organizing. The second researcher's goal was to understand the spontaneous and individual practices of each team member *in situ*. This involved understanding what an actor was doing both objectively and subjectively at a given moment, whether he or she was pitching a tent, progressing for an hour, or taking part in a preparatory meeting.

The second researcher was integrated into the group. Unlike the first researcher, the second researcher was not a member of the team but a novice to polar ski expeditions, an individual who sought to learn the practices of each team member, an observer who followed the group in its path, and who contributed to all collective tasks without assuming any responsibility or decisions. Compared to the first researcher's observing participation, the second researcher's role can be described as participant observation. In fact, his role was not to intervene in the actors' practices, but to investigate them first and foremost (Peretz, 2004). Participating in the group's activity by following it to understand what is happening is not intended to recover a "given reality" in a neutral and comprehensive manner. As Favret-Saada (1977) points out, even would-be external observers participate in a studied situation through the way in which they watch the group and through the way in which the group watches them. We therefore invert the terms to describe a situation in which the second researcher created greater distance from the group *in situ*. This role gave the researcher more time to observe, to take notes, and to implement various systems of investigation *in situ,* and it spared the researcher from having to stake a position in relation to the other members of the group. This way, the second researcher could share the intimacy of each member's experience *a posteriori* without making them feel judged or being compared. Accordingly, after the dogs were lost during the expedition, the first researcher not only expressed himself regarding the severity of the event but also proposed and implemented a makeshift solution. The second researcher, however, maintained a degree of passivity and instead focused on concealing his own feelings and emotions, which in their own way also constituted a form of judgment.

During his participant observation, the second researcher could use his relative distance from the group to gather various observation materials regarding the practices of each team member. However, to understand these practices, a researcher must go beyond what is observable and approach the action's implicit, personal, and significative aspects, that is, the way in which each actor lives his or her situation and acts in a given moment. Insofar as an actor's way of being, living, and acting in a particular context is expressed, above all, in terms of actions and remains largely prereflected, a system must be used to encourage and help the actor make his or her practice explicit. To this end, we turned to a new methodology: the *subjective re situ* interview (Rix & Biache, 2004; Rix & Lièvre, 2008, Rix-Lièvre, 2010). Inspired primarily by self-confrontation (Theureau, 1992) and the explicitation interview (Vermersch, 1996), this methodology focuses on the prereflected aspects of action by simultaneously mobilizing both observations and materials related to the subjectivity of the actor being studied. It has three precise objectives. The first is to focus an actor on a moment chosen by the researcher and try to have the actor make explicit the experience of that moment. The second involves linking this explicit experience to an act that has already been performed and documented, which allows a researcher to control the link between discourse and an act and also to document the act itself using various types of materials. The third objective is to remain as close as possible to an actor's way of being *in situ*.

In concrete terms, the system implemented for each moment investigated *in situ* involves filming:

1. The behavior of an actor within his or her context
2. A point-of-view shot close to that of the actor in action and taken by a head-mounted camera, which is known as a *situated subjective* perspective

The actors are then questioned one by one during *subjective re situ* interviews *a posteriori* (e.g., after a meeting, at technical weekend, or a preparatory trek). Each interview is conducted by a researcher, who must bring an actor back to a specific moment using the record of his *situated subjective* perspective, while encouraging the actor to make his or her way of experiencing that moment explicit. The *subjective re situ* interview therefore is a tool that a researcher can use to approach the individual experience of each actor at various moments. A researcher is able to enter each actor's world of practice, one by one and *a posteriori,* as well as to maintain a distance from the group and its activities *in situ.* The approach constructed during participant observation is vital: In order to grasp the significative side of each actor's practices, a researcher must construct the possibility of sharing experience throughout the course of field work. The essential preliminary steps for our investigation were

- The voluntary participation of each team member
- Their acceptance of both the research and the researcher*
- The construction of a trust-based relationship† between each actor and the researcher

---

* Accommodating research and the presence of a researcher has rarely been difficult since the inception of traditional polar expedition research.

† To create a trust-based relationship, the researcher must avoid taking a position of superiority or judging the decisions of members and/or the group. He or she must avoid being identified as a subgroup or a privileged ally of one person or another and become a participant who can listen to the actors describe their experiences without social censure.

As previously discussed, this is why we paid particular attention to the researcher's role within the group.

Finally, we could document each investigated moment through the simultaneous use of an actor's verbal statements, which provide a subjective view regarding his or her experience, along with two recordings made *in situ*: one for behavior and one for the *situated subjective* perspective. These different types of materials—fairly subjective on one hand and fairly objective on the other—could be used jointly to reconstruct the process surrounding an action. This way, we could define and understand each team member's specific, spontaneous, and situated practices.

5.2.3. How Should the Two Research Roles Complement Each Other to Better Understand Organizing in Situ*?*

This section discusses the relevance of the dual investigation method for the polar ski expedition's organizing. According to positivist epistemology, placing two researchers in the same fieldwork to study the same group of actors makes little sense beyond corroborating the validity of observations. From a constructivist point of view (Le Moigne, 1995), however, this approach provides an opportunity to develop two perspectives that differ on several levels within a single research project:

1. Novice-Expert
2. Actor-Follower
3. Individual-Collective

The first researcher's account highlights the collective's functioning when everyone involved agrees on the reconstructed course of the expedition. In parallel, the materials constructed by the second researcher help us understand the practices of each actor, as well as their motives *in situ*. The proposed observatory therefore places a researcher's relation to fieldwork, or more precisely, his or her relation to actors, at the center of the material construction. The fact that all researchers participate in a studied situation no longer appears to create bias, but rather provides an opportunity to construct various research materials (Berry, 2000; Favret-Saada, 1977; Girin, 1990; Plane, 2000). Each researcher's role provides an opportunity to document the various facets of organizing through the researcher's specific interactions with the actors. A researcher's role as an instrument for constructing material is implied through his or her interactions with the actors. Here, the lessons garnered from Favret-Saada (1977; 2009) acquire their full meaning, because the researchers' exchanges and interactions differ in nature according to their respective roles and positions. Fieldwork, and participant observation in a broader sense, therefore, provides a possibility of constructing a particular approach based on the way in which actors perceive a researcher. As a result, construction of materials does not depend on researchers' adopted role but on their way of interacting with actors, which, in turn, is largely dependent on the position conferred on them by the actors. The proposed roles should not be applied mechanically—because fieldwork is always fluid—but should instead provide reference points for researchers. In fact, if a researcher accepts "to be affected" (Favret-Saada, 2009, p. 158), that is, to be submerged in the human experience under observation—which was inevitable during a polar expedition where the researchers' physical integrity was at stake—these reference points can be used to direct a researcher's interactions with others (Rix-Lièvre & Lièvre, 2014). They become all the more significant with regard to the simultaneous presence of two researchers throughout the course of a single fieldwork, where indeterminate roles could lead to the construction of similar materials and substantially decrease the effectiveness of their dual presence.

The two roles therefore appeared to be complementary for understanding the polar expedition as both a project and organizing (but by no means for understanding a given reality in its totality). If the activities of a collective cannot be reduced to the sum of its individual practices, understanding how a group self-organizes is impossible when we limit ourselves to studying these practices individually. Inversely, approaching the genesis of a collective activity, or that which stems from interaction, seems impossible without considering the actions of each actor along with his or her cognitive, affective, and tangible motives. Only by echoing these two can we begin to understand organizing.

5.3. Conclusion

We constructed this observatory technique over several years of experimentation and fully integrated it within the framework of a 2005 polar expedition. We monitored this project in its entirety for two years, and our observation directly addresses issues surrounding the investigation of collective practice within projects. The methodological problems regarding the investigation of practices *in situ* are largely underestimated by management sciences. Only by investigating the research of anthropologists, work psychologists, and ergonomists were we able to construct this tool. Paradoxically, the chosen field of polar expeditions made it easier to resolve the methodological problems of acceptance, secrecy, and privacy, as discussed by Midler (1996).

This system is in line with the Weickienne perspective on organization. From a theoretical standpoint, we believe that defining organizing in its full actualization could rest on a "meta" level that includes both the logic of the collective actions documented by the MLB and the logic of the individual actions documented by the formalizations. This theoretical approach leads to a specific understanding of organizing. For instance, the knowledge gained regarding polar bear risk management, based on records from both the MLB and SPOS, does not match the knowledge gained from either record taken separately. Using the MLB account, we may draw conclusions regarding the expedition leader's not properly managing the dogs or bear risks. On the other hand, the formalizations from the SPOS showed that security systems mattered to the expedition leader only when bears were present.

The issues surrounding the observatory's cumbersome nature should be considered against the specific qualitative method used, which focuses on "ways of doing" rather than "ways of saying." The resulting challenge for the "project as practice" approach is to find investigation tools that can handle issues of actual practice. In fact, the observatory allows us to study organizational dynamics involved in collective activity by using different types of materials, including the observation of activities *in situ,* as well as the more or less formal verbalizations of actors. This way, the observatory avoids reducing the study of activities to the study of activity-related discourse and accounts for the way in which actors experience and describe their situations. The advantages and limits of this observatory should therefore be gauged according to what it does reveal: the conditions under which organizing occurs.

To conclude, we must address the means by which such methods can be applied to more classical management situations. This particular tool, developed in a context of polar expeditions, clearly has utility when investigating typical collective management practices, more classical organizing practices, and, especially, the technical problems associated with cold and snow. However, this observatory is intrusive, and getting actors to accept such an intrusion into their everyday working life requires considerable preparatory work along with the building of trust between researchers and actors. This trust is not a given and must be developed over time. Although the system partially transforms a situation, it must never disturb it; the activity must remain at the forefront. How the materials obtained from organizing could be used must also be explained to actors at the outset. In fact, delivering the materials obtained from this type of observatory to the actors represents a significant step in terms of transparency in various management situations, thereby avoiding weighty power-based and strategic issues.

References

Berry, M. (2000). Logique de connaissance et logique d'action. *Cahier de recherche de l'Essca, 7,* 3–58.

Blomquist, T., Hällgren, M., Nilsson, A., & Söderholm, A. (2010). Project-as-practice: In search of project management research that matters. *Project Management Journal, 41*(1), 5-16.
Bourdieu, P. (1977). Outline of a theory of practice. *Cambridge Studies in Social Anthropology*, (16).
David, A., Hatchuel, A., & Laufer, R. (2000). *Les nouvelles fondations des sciences de gestion* (pp. 83–108). Paris: Vuibert.
Dosse, F. (1995). *L'empire du sens. L'humanisation des sciences humaines.* Paris: La Découverte.
Dubar, C. (1991). *La socialisation: Construction des identités sociales et professionnelles.* Paris: Armand Colin.
Favret-Saada, J. (1977). *Les mots, la mort, les sorts.* Paris: Gallimard.
Favret-Saada, J. (2009). Désorceler, Paris. *Éditions de l'Olivier.*
Garel, G. (2003). *Le management de projet.* Paris: La Découverte.
Garfinkel, H. (1967). *Studies in ethnomethodology.* Englewood Cliffs, NJ: Prentice Hall.
Ghasarian, C. (2004). *De l'ethnographie à l'anthropologie réflexive. Nouveaux terrains, nouvelles pratiques, nouveaux enjeux.* Paris: Armand Colin.
Girin, J. (1990). L'analyse empirique des situations de gestion. In A.-C. Martinet (Ed.), *Epistémologie des sciences de gestion* (pp. 141–182). Paris: Economica.
Hlady Rispal, M. (2002). *La méthode des cas. Application à la recherche en gestion.* Brussels: De Boeck Université.
Lahire, B. (2002). *Portraits sociologiques. Dispositions et variations individuelles.* Paris: Nathan.
Le Moigne, J.-L. (1995). *Les épistémologies constructivistes.* Paris: PUF.
Lévi-Strauss, C. (1950). Introduction à l'œuvre de Marcel Mauss. In M. Mauss, *Sociologie et anthropologie* (pp. IX–LII). Paris: PUF.
Malinowski, B. (1963). *Les Argonautes du Pacifique occidental.* Paris: Gallimard.
Midler, C. (1996). *L'auto qui n'existait pas.* Paris: Dunod.
Peretz, H. (2004). *Les méthodes en sociologie. L'observation.* Paris: La Découverte.
Plane, J.-M. (2000). *Méthode de recherche-intervention en management.* Paris: L'Harmattan.
Quéré, L. (1993). A-t-on vraiment besoin de la notion de convention? *Réseaux, 11*(62), 19–42.
Ricœur, P. (1983). *Temps et récit. Tome 1: L'intrigue et le récit historique.* Paris: Seuil.
Rix, G., & Biache, M.-J. (2004). Enregistrement en perspective *subjective située* et entretien en *re situ subjectif*: Une méthodologie de constitution de l'expérience. *Intellectica, 38,* 363–396.
Rix, G., & Lièvre, P. (2008). Towards a codification of practical knowledge. *Knowledge Management Research and Practice, 6,* 225–232.
Rix-Lièvre, G. (2010). Différents modes de confrontation à des traces de sa propre activité. Entre convergences et spécificités. *Revue d'anthropologie des connaissances, 4*(2010/2), 357–376.
Rix-Lièvre, G., & Lièvre, P. (2009). L'observatoire de l'organisant: Mode d'interprétation

des matériaux qui en sont issus. *Revue Internationale de Psychosociologie,* Numéro spécial sous la direction de Martine Hlady Rispal, 15, 161–178.

Rix-Lièvre, G., & Lièvre, P. (2014). Rôle d'un dispositif d'investigation posé a priori dans l'exercice d'une réflexivité méthodologique. La petite histoire de l'ethnographie d'une expédition polaire à ski. *Recherches qualitatives, 33*(1), 149–171.

Rouleau, L. (2007). *Théorie des organisations.* Montréal: Presses universitaires du Quebec.

Soulé, B. (2007). Observation participante ou participation observante? Usages et justifications de la notion de participation observante en sciences sociales. *Recherches qualitatives, 27*(1), 127–140.

Theureau, J. (1992). *Le cours d'action, analyse sémiologique: Essais d'une anthropologie cognitive située.* Berne: Peter Lang.

Theureau, J. (2000). Anthropologie cognitive et analyse des compétences. In Centre de recherche sur la formation du conservatoire national des arts et métiers (Ed.) *L'analyse de la singularité de l'action* (pp. 171–211). Paris: PUF.

Vermersch, P. (1996). *L'entretien d'explicitation.* Paris: ESF.

Vermersch, P. (1999). Pour une psychologie phénoménologique. *Psychologie française, 44*(1), 7–18.

Vermersch, P. (2004). Prendre en compte la phénoménalité: Propositions pour une psycho phénoménologie. *Expliciter, 57,* 35–45.

Weick, K.E. (1979). *The social psychology of organizing.* New York: McGraw-Hill.

Weick, K.E. (2003). Préface. In B. Vidaillet, *Le sens de l'action.* Paris: Vuibert.

Chapter 6

A Traditional Cree Expedition on the Ancestral Lands of the Neeposh Family of Northern Québec

Nathalie Guérard and Anne-Marie Cabana

Northern Québec features several great high-flow rivers, which makes this area prime for hydroelectric development. Hydro-Québec's and the Société d'énergie de la Baie-James's (SEBJ) construction of the Eastmain 1-A and Sarcelle generating stations and the diversion of the Rupert River would soon cause the flooding of an area northeast of the diversion point. This flooding, planned for November 2009, would submerge part of the trapping grounds used by generations of a Cree family, the Neeposh.

In accordance with federal and provincial laws, the Rupert River diversion project was the subject of an environmental impact assessment between 2002 and 2006. Accordingly, the consulting firm GENIVAR was tasked by Hydro-Québec with characterizing the fish populations in the areas affected by the project, including lakes within the Neeposh family's territory.

The fish population study was conducted by selecting sampling sites that were representative of the different areas affected by the project, with experimental fishing following well-established criteria and protocol. However, the results were contested on numerous occasions by members of the Neeposh family, who claimed that the description of the fish communities included in the impact assessment did not correspond to their view of the situation.

Indeed, the project sponsors were confronted with a problem in the study's methodology. The scientific techniques used to characterize the fish and the Neeposh family's traditional approach had two different aims. Integrating the two was going to require much more flexibility than first thought.

The protocol developed to characterize the fish populations was based on the scientific method that uses objective and reproducible techniques. This was used to randomly record the fish species present in a given area, calculate their relative abundance based on fish catches, and describe their distribution. The object was to obtain in a few weeks a detailed description of the fish communities and to evaluate the average fishing harvest of the area.

On the other hand, the Cree use traditional methods for subsistence fishing. The Cree target species that are best for eating and choose to fish the most productive sites for these fish. Also, their gill nets do not have the same characteristics as those used for scientific inventories. This inevitably leads to different results when comparing the portrait of the fish communities obtained by each method.

In short, traditional knowledge takes into account site accessibility, distance from camps and trapping sites, weather conditions and their variability over the years, and the prized species. The time scale and perspective are completely different and even sometimes contradictory.

During discussions conducted with representatives from the SEJB, Hydro-Québec, and GENIVAR, members of the Neeposh family expressed their desire to study the distribution and population of different fish species using traditional Cree capture methods, focusing on brook trout, which is a species they value. The Neeposh stated that this fish was more abundant in their territory than the scientific survey seemed to show. This was why they requested that the survey be redone at sample sites of their choosing and using their own gill nets. They wished to compare the scientific method and the Cree method.

Members of the Neeposh family agreed with the SEBJ and Hydro-Québec that a team of observers from GENIVAR would accompany them during their traditional fall fishing expedition in 2006 so they could better understand their viewpoint and knowledge of the area. GENIVAR's role was to record and produce an account of the different activities carried out during this expedition.

This chapter summarizes the experience of this traditional fishing expedition. No attempt is made here to address the anthropological aspects of Cree tradition and knowledge. Rather, we relate how, in the context of an environmental impact assessment, the sponsors of a hydroelectric project attempted to adapt a formal fish population characterization process to the viewpoint of a Cree family, the Neeposhes. Integrating their knowledge of the study area was made possible by experiencing a traditional fishing trip on their ancestral lands. It is this exceptional experience that we wish to share. This chapter is in large part an adaptation of the journal that was updated daily by the principal author of this chapter, who accompanied the Neeposh family on this trip. Information regarding Cree traditions is taken mostly from conversations with trip participants.

6.1. Project Organization and Logistics

Although a lot of trapping and hunting took place during the trip, the project was focused mostly on the search for brook trout. The Neeposh family planned the trip so that it would cover most of the fishing sites known for brook trout. This route also partially followed the one traditionally used by Neeposh between the village of Mistissini and their winter homes.

The project participants included several members of the Neeposh family, members of the Mistissini community related to or affiliated with the Neeposhes, as well as two GENIVAR wildlife technicians (see Figure 6.1). The Neeposh family formed four fishing teams of two to four people. The Neeposh and Mistissini community members spanned three generations, bringing their expertise, wisdom, experience, and dedication to ensure that the project ran smoothly.

This project required lengthy and meticulous preparation. It is Cree habit and custom to have every decision reviewed by the family. Family members therefore held several preparatory meetings among themselves and with representatives of GENIVAR and the SEBJ.

All the necessities for living in the forest and long canoe trips were gathered and carefully packed by the team members. The basic equipment included clothing, canoes, outboard motors, firearms, fishing gear, and various traps for fur-bearing animals. In addition, there was camping equipment, cooking utensils, communication devices, security equipment, first-aid kits, camp stoves, etc. A small mobile laboratory was also brought to take measurements and record observations (e.g., length, weight, and sex) of the main species of fish caught. Finally, there were, of course, food provisions. Even if most of the food eaten was to be the fish and game caught during the trip, a considerable amount of basic foodstuffs was purchased to meet daily needs.

It has to be pointed out that the equipment used by the Cree was not at all the usual high-tech equipment used in an adventure expedition. It was typically not compact, light, or modern. During the journey, team members had to carry the equipment when faced with rapids that the canoes could not cross. This required several trips back and forth along narrow forest paths. In fact, a

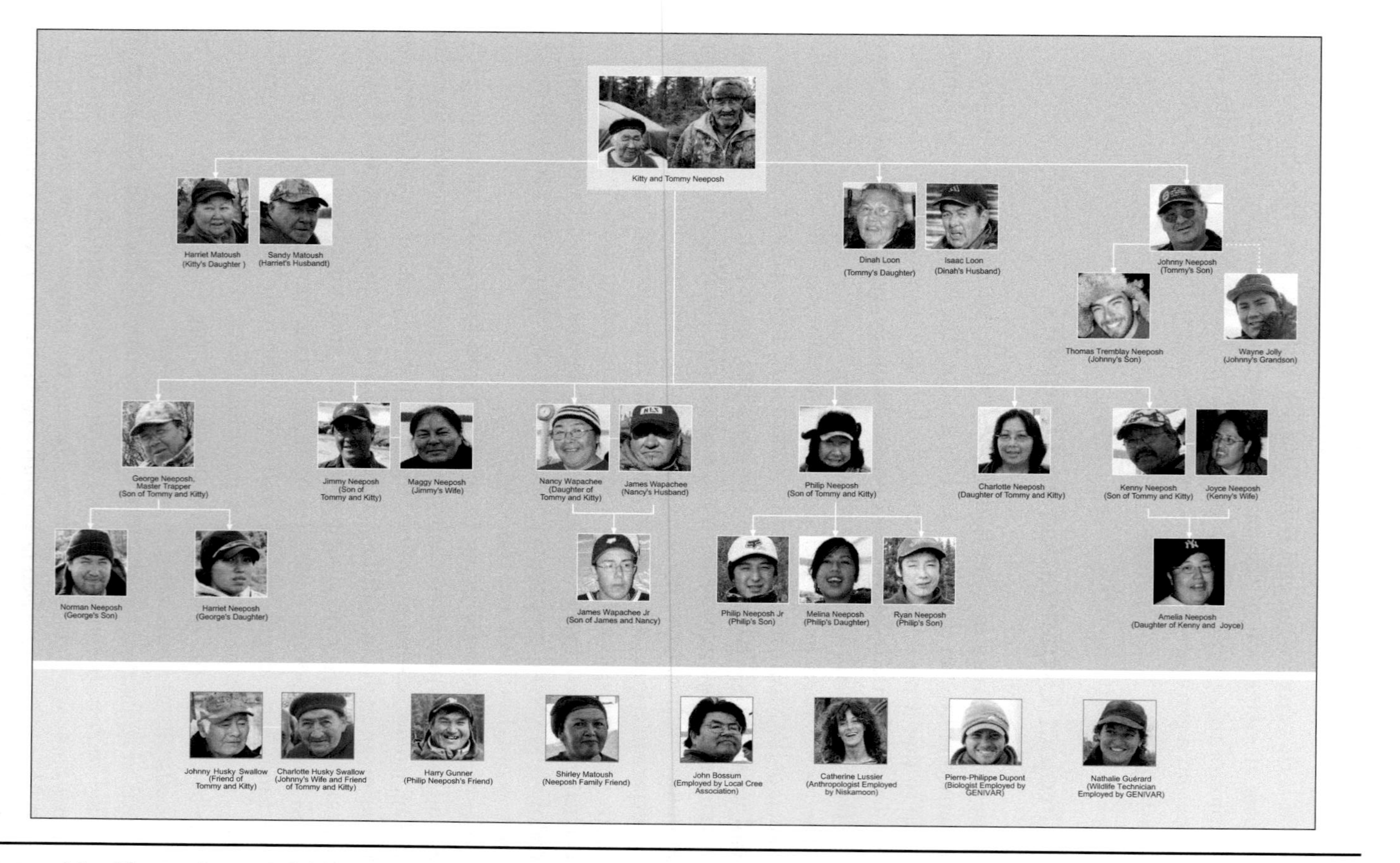

Figure 6.1 The project participants.

helicopter had to be used a few times when portage over land or canoeing would have taken too long for the time allotted to the project or when water levels were too low for the canoes. The Cree in turn needed to be flexible regarding not always using the traditional methods of transportation.

6.2. The Trip and Camps

This 8-week expedition (from mid-September to mid-November) departed from a point located near the Rupert River, at the confluence with the Mistikw Awaashsiipii (Misticawissich) River. The trip was to follow a chain of lakes and waterways northward over 50 km, ending at the permanent Neeposh camp on the shores of Lake Kaa Paschchisheyaau (Deschamps). Nine other temporary camp sites were set up near fishing sites during the trip.

This trip, traditionally taken by the Neeposh family, required numerous portages to avoid several stretches of rapids too difficult to cross by canoe. These portages required a lot of time and energy. Just imagine what effort is required to transport a 6-m-long canoe along with firearms, equipment, and young children, over more than a kilometer.

For safety and cooperation reasons, two to five families usually traveled together to reach the Neeposh territory. They would leave the village of Mistissini in mid-August and travel for almost a month to reach the camp at Lake Kaa Paschchisheyaau. The trappers would then disperse with their families to hunt and trap in various locations. After the winter season, each trapper would return to Mistissini with an average of roughly 100 pelts. The main species sought were beaver, otter, mink, American sable, muskrat, fox, lynx, and American hare. Selling these pelts was their only source of income, and the trappers would return to Mistissini around mid-June to trade them. The two summer months were their leisure time and also their time to be with other members of the community. They also would work for the community in exchange for such commodities as flour, sugar, rice, oats, bacon, and tea, in lieu of salary. Loaded with several hundred pounds of these provisions, they would return in mid-August for another year of life in the wild.

6.2.1. Welcome Feast

The first day of this amazing adventure began with a welcome feast prepared according to Cree tradition and during which Mr. Tommy Neeposh, the family patriarch, gave a moving speech. He expressed his wish to see all project participants work with heart, joy, and peace.

Tommy died at the age of 97, about two years after this adventure. He led an extraordinary life and had witnessed many periods of the history of the Cree of James Bay. During his speech, he looked back on his life, his land, and what he had taught his children, including respecting the Earth and all people. He said he always knew that big changes would come and that the Cree would have to adapt to a modern way of life that is constantly evolving. He wished with all his heart that his children and grandchildren would grow up and accept this change in the hopes of a better future for future generations.

Although the division of their land for the benefit of a hydroelectric project was an emotionally charged issue, the family members accepted Tommy's decision out of respect for this wise man.

6.2.2. Setup of a Typical Campsite

The first camp used along the Rupert River belonged to George Neeposh, the tallyman who succeeded his father, Tommy. This is a hunting camp that has a wood shelter and a wood stove. Although it was a permanent shelter, it did require some arranging to ensure the teams were comfortable during their stay, which was to last about a week. A tent had to be put up near the camp to accommodate all the team members. Each team member lent a hand with the setup, which they all knew how to accomplish.

In general, a camp's finish and level of comfort are a function of the length of the stay in the camp. Each type of camp had a specific use. Permanent camps usually consist of a wooden structure or a shelter covered with canvas. Such camps are found in favourite locations near fishing and trapping sites often used in the fall and winter. Temporary camps, which are more rudimentary, consist of tents or teepees that are used as shelters for short periods of time. These camps are taken down at the end of each stay, and nothing is left behind except for the poles to be used at a later date. They piled these poles carefully by standing them up on a tree so not to ruin them. These campsites are situated along the usual trapping routes and trails used by the Cree. Although the number of these sites has decreased over the years, the exact locations of these former camps is etched into the memory of those who used them. At the time when members of the Neeposh family used to travel by canoe between Mistissini and their winter camp, they also set up temporary camps at the start and end of long portages, which sometimes lasted more than a day.

Whether a campsite is permanent or temporary, its location is always chosen carefully. The selection of a site is guided largely by the availability of resources nearby and abundance of pine branches used to prepare the tent floors. This allows the camps to be easily reused during the winter without having to travel long distances. The selection is also influenced by the proximity to transportation routes. In winter the Cree prefer to move away from rivers that usually do not freeze over enough and can be dangerous to follow. Lakes are safer and are often home to beavers whose fur is sought after by the Cree. Therefore, by placing a camp in an area where there are many small lakes, the Cree improve their chances of trapping beavers.

In the spring, it is preferable to return close to the rivers in order to access Canada geese migration areas after the spring thaw. The camp at Lake Kaa Paschchisheyaau is in a central location for this. George Neeposh's camp, for its part, is located in a strategic spot for the first part of the journey, given the proximity to the brook trout fishing sites that were the subject of the study.

6.2.3. A Typical Day

The team members started their day at the crack of dawn. However, with the increasing cold of September mornings, team members sometimes stayed in their sleeping bags and waited for the tent to be warmed up by a fire started by a hardier teammate. The Cree have a great sense of humor that made waking up a special time marked by laughter and fun.

Once washed, each member became busy with a task. Often the women prepared breakfast while the men made sure the boats were filled with gas, and checked the fishing, hunting, and trapping equipment. Each gathered up his or her personal gear: rain clothes, warm change of clothes, a thermos of tea, and snacks.

Following tradition, the men must cut a large amount of dry firewood to make sure the camp does not run out too quickly. Gathering quality firewood is an essential task for the group's well-being and is considered a priority activity. Black spruce is the preferred species, and their dried stumps are sought after. Sometimes fresh wood was collected because it was slow-burning for chilly nights. At George Neeposh's camp, this job was made easier because a forest fire had ravaged the area in 2002 and left behind many tree stumps.

Whether it was collecting firewood, setting up camp, or fishing, hunting, or trapping activities, these tasks required a significant number of operations, from planning to execution, up to the preparation of the captures. In such a situation, it was difficult, nigh impossible, to keep to the initial work plan. The team members had to know how to balance the study's requirements with the human requirements, specifically those related to family life and the Cree culture. Rigor and flexibility were thus continually in conflict during the expedition. Despite the traditional aspect of this project led by the Neeposhes, rigor was required because the fishing had to follow a minimal protocol to meet the project's objectives. Flexibility was also required because on such an adventure, life follows nature's rhythm, in a nonconventional way where tradition has precedence over science.

When travelling by boat, the team members were always on the lookout for forest animals. Thus, when the teams left for the fishing sites, the day's plan could be completely thrown off by a bear or moose sighting. The team leaders would decide spontaneously to go after the animal, tracking it, sometimes all day until nightfall.

These chance meetings with animals were the topic of evening's discussions. Over a good meal of beaver or another type of game roasted on a stick, the next day was planned and tasks assigned. For example, a team would be given the task of setting bear or beaver traps, while other team members would be in charge of looking for rapids suitable for angling.

Once the meal was over and the dishes put away, each person would snuggle in his or her sleeping bag to listen to the storytellers until nightfall. The families, their friends, and the wildlife technicians all slept under the same roof. When all was quiet, the GENIVAR observers wrote up their accounts of the day by the light of their head lamps.

A few photographs are included in this chapter to illustrate some of the typical activities carried out during the expedition. They describe the traditional way of cooking fish (on a *bonask*, see Figure 6.2), the method for portaging canoes and material (see Figures 6.3–6.6), and the technique for skinning beaver (see Figures 6.7 and 6.8).

Figure 6.2 Cooking on a *bonask*.

Cooking on a *Bonask*

The Cree term *bonask* means a stick that has been cut to a point and had its bark peeled off. It is used to cook food on a fire. One skewers fish or meat on it as well as rolling bannock (a Native American Indian bread, also called traveler's bread). This way of cooking is simple and delicious. Walleye or lake trout, for example, are first gutted and then pierced with the stick from the eye to the ventral fin. They are then tied up with cotton string which has been soaked in water to keep the fish from falling apart during cooking. They are then planted near the fire and turned regularly (see Figure 6.2).

Navigation and Portage Methods

Several techniques are used to navigate around obstacles. Sometimes it is necessary to transport all the equipment and supplies along a path. Brush must first be cut away to clear a path, and then the canoe is dragged onto dry land and emptied of its cargo. The travelers transport their bundles with the help of a leather strap that passes across the carrier's forehead to support a part of the load (see Figure 6.3). To transport a canoe or a rowboat, one places the paddles perpendicularly across the ribs or on the seats and ties them with a strap. This

Figure 6.3 Carrying cargo with a forehead strap.

Figure 6.4 Two-man portage of a canoe.

Figure 6.5 Portaging a canoe by pulling it.

Figure 6.6 Towing and steering a canoe.

distributes the weight of the boat or canoe, which one lifts and carries upside down on the shoulders. A second person can help by guiding the prow (see Figure 6.4). If the distance to cover is not too long and the ground surface allows it, one can simply pull the boat with the help of ropes and handles (see Figure 6.5). Portage is not always needed to cross rapids. Sometimes a tow line is all that is needed. The boat is towed by a line held by one or two people on the bank. A person can also stand in the boat and help guide it by using a long pole (see Figure 6.6).

Skinning Beaver

Preparing pelts, from skinning the animal to drying the pelts, is a delicate job that is usually performed by women and requires great care to avoid spoiling the fur, which decreases their value. One starts by laying the animal on its back, then, after finding the chest bone, one traces with a finger a straight line from the animal's chin to its tail. A ventral cut is made along this line and care is taken not to pierce the flesh; then the skin around the paws and tail is cut (see Figure 6.7).

Beginning at the jawline, one side at a time, the fur is separated from the skin with the help of a knife with a curved blade. This is done carefully to avoid

Figure 6.7 Cutting a beaver pelt.

Figure 6.8 Scraping fat from a pelt.

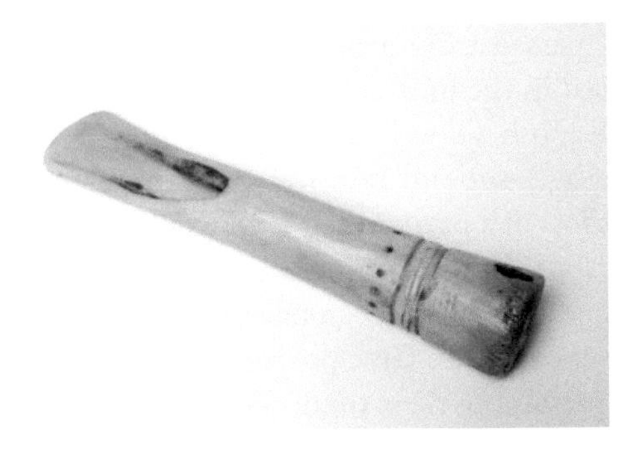

Figure 6.9 Caribou bone scraper.

piercing a vein or artery and dirtying the fur with blood. For the fattier parts of the animal, one uses a sharp scraping knife (see Figure 6.8) that was carved from a caribou's femur and beveled (see Figure 6.9). Pulling the fur against the thigh, one quickly moves the scraping knife from top to bottom to separate the flesh from the fur.

Once the pelt is pulled off, one scrapes the inside to remove any flesh and fat and then washes it in soapy water to make the fur shine. To dry the pelt, it is either stretched and tied to a hoop, usually made of birch or pine, or pulled and nailed to a plywood board.

6.3. Assessment of the Journey

For many, the loss of a land and a traditional way of life are relatively abstract notions. The time spent accompanying a Cree family during a traditional trapping expedition provided us with a better understanding of this reality. During this journey, the observers were able to see just how attached the Neeposh are to their land, of which they have a deep understanding and knowledge. Every river bend, every portage, every hill evokes anecdotes, memories, and stories, sometimes sad, sometimes happy. Indeed, this ancestral land is where the family's collective memory is written, passed from generation to generation.

It is in this land that most of them were born and developed, over the years, a knowledge of the land, the animal species, and the principles of survival. It is also here that they acquired their social and spiritual values, forming the basis of their existence. Although the Cree no longer live exclusively by fishing, hunting, and trapping, these activities are tied to a way of life that is the core of their culture and the main source of their traditional knowledge. And it is the value of this way of life, passed from generation to generation, that the Neeposhes want to see addressed in the impact assessment.

Mr. Tommy Neeposh, as well as the members of his family, consented to their land undergoing profound transformation, despite the sacrifice it represents for them. The loss of a part of their land would undeniably have an impact on their way of life, but they accept to share their land so that their children, as well as all of the people of Québec, can have a better future. They wish from the bottom of their hearts that the Cree people can benefit from this development project, while still preserving their traditional values.

In recognition of the vision and spirit of sharing that Mr. Tommy Neeposh showed during Eastman1-A and la Sarcelle powerhouses and Rupert diversion project, SENJ and Hydro-Québec named the transfer tunnel that channels the water diverted from the Rupert River toward the La Grande hydroelectric complex after him. They also erected, with the collaboration of the Neeposh family, a commemorative scenic outlook at the upstream entrance of the tunnel.

Chapter 7

Borrowing Concepts from Expedition Travel to Stimulate Alternative Tourism

Alain A. Grenier

The sinking of the cruise ship *Explorer* in Antarctica on November 23, 2007, might well have sounded the death knell for cruise tourism on the seventh continent. Almost 30 years earlier, in 1979, the crash of Air New Zealand flight 901 over Antarctica, killing all 257 passengers and crewmembers, had temporarily halted tourist sightseeing flights over the frozen continent. Yet the loss of the *Explorer* actually spurred demand instead of killing interest in polar expeditions. Why?

For almost three decades, the environmental crisis has been driving ecological awareness and inevitably promoting natural areas, especially spectacular ecosystems that feature outstanding entertaining locations for the general public. The response has been so great that industry observers have been talking about the advent of mass nature tourism. It is no surprise that operators are striving to diversify their tourism options in natural settings: from crossing oceans to hiking in polar regions. Soon, there could be even outer space, pushing adventure tourism to the limit. These experiences may be private or commercial, such as a unique tour experience in an area still fairly undeveloped by operators. Given the wide range of packages now available, operators are increasingly using the term "expedition" to make their products stand out, much to the annoyance of old adventure hands who fear that the image long associated with exploration is being diminished.

Describing an organized trip as an "expedition" may very well serve the tourism industry's interests, given the quest for distinction (Boyer, 1995) and more physically and intellectually challenging products. From a social management perspective, however, the term "expedition" raises various issues: organizational (safety, support, and rescue); economic (who should pay for rescue operations); and legal (does national sovereignty create an obligation to rescue all adventurers imperiled within a country's boundaries?). In environmental terms, do extreme sports or expeditions in natural areas promote an experience in harmony with the environment or a harmful conquest? The merging of expeditions and tourism, as well as the ever more massive numbers of tourists visiting natural sites on so-called adventure or expedition travel, simply increases the seriousness and urgency of finding answers to these questions.

Based on a literature review, this chapter strives to understand the nature and implications of linking the concepts of expedition and tourism. I survey the concepts borrowed by the adventure tourism industry from expedition travel and we question use of the word "expedition" as an enticement for tourists, and its consequences.

7.1. Tourism and the Environment

The ecological awareness that followed the environmental crises of the 1970s, combined with criticisms of mass tourism's negative impact on natural ecosystems and cultures, has stimulated the growth of alternative tourism to natural environments. Consistent with the ecological thinking of the 1980s, this alternative has been shaped by the concept of ecological tourism—the so-called "ecotourism." This ecotourism focused on spectacle ecosystems—very unusual ecosystems that provide outstanding, entertaining features. Their remoteness both protects the environment and rewards tourists unconsciously (or not) seeking distinctiveness (Bourdieu, 1979). Distinction has indeed always been the driving force in tourism (Boyer, 1995), which explains in large part tourists' quest for unexpected and new sensations. Wrongly perceived as the opposite of mass tourism, ecotourism therefore provided an alternative to the ennui of the ordinary (Lee & Crompton, 1992). Only ecotourism—an environmental approach to tourism management—was not so much ecological as increasingly mass-oriented nature-based tourism.

Spurred over the past three decades by environmental discourse, tourism in natural settings has grown so extensively all around the world that we now refer to its "massification" (Grenier, 2009, p. 18; Clifton and Benson, 2006, p. 238; Butcher, 2005, p. 114; Diamantis, 1999, pp. 93, 116; Acott, La Trobe, & Howard, 1998, p. 239; Burton, 1998). While the negative impact of mass tourism can be contained in urban areas, a natural environment increasingly suffers damage from the repeated presence of tourists. Tourism's negative environmental impact is caused primarily by overconcentration of visitors (i.e., saturation of carrying capacity) combined with inappropriate behavior toward the ecosystem (e.g., soil compaction, erosion, and repeated disturbance and harassment of animal species). Some of this deleterious behavior can be attributed to visitors' ignorance, which requires guides and educational programs. Most of this behavior is linked to the attitude of a certain class of tourist known as "conquerors" (Grenier, 1998; Viken, 1995). This type of visitor often places priority on attainment of personal goals (e.g., geophysical and mental) to the detriment of environmental conservation. In the Galapagos, for example, ecotourism, once considered a boon, is now criticized and identified as a threat to the integrity of this ecosystem (Basset, 2009; D'Orso, 2002). Although managers of parks and other natural environments visited by tourists have adopted management approaches designed to reduce and minimize the traces left by masses of visitors to the natural environment, there is no denying that many touristic ecosystems have deteriorated, and management methods currently in place to handle the growing flood of tourists have limited effect. The industry apparently cannot grasp the concept that whether a travel experience is ecological depends not on the nature of the *destination* but rather on user *behavior* (Grenier, 2009, p. 18).

In studies of visitors to polar regions, Grenier (1998, 2004) notes two main motives of tourists in natural environments: those focused on harmony with nature and those focused on conquest and domination of the environment. Tourism offerings can actually be classified by the values they promote: harmony or conquest and domination. In turn, each orientation dictates a different management model. In the first case, preserving the integrity of the site visited is the core concern of managers and participants. Visitors must behave in a manner respectful of nature and limit their actions and access. This biocentric approach makes environmental integrity the central experience. By contrast, a conqueror's goal is to reach a specific place that is very difficult to access, usually due to challenging terrain or climate (see Figure 7.1), for the purpose of earning a social reward (in the eyes of peers, at least). This visitor often must deploy substantial technical and logistical resources to attain the goal. This is the egocentric facet of nature tourism. Renewal of nature tourism through conversion to the adventure and expedition approach should worry managers concerned not only with public safety but also conservation of the environment.

Figure 7.1 Obstacles such as ice, which obstruct access and make success of a trip uncertain, help create a sense of adventure. (Photo by Alain A. Grenier.)

7.2. From Tourism to Adventure

To some extent, tourism and adventure have always been intertwined. From the days of the Grand Tour (circa late 1600 to early days of 1840 and large-scale rail transit) to modern-day tourists, those who choose to venture away from home have always strived for the exotic and to seek the unfamiliar. In the early days of leisure travel, tourists required the services of guides, translators, and the like to soften the shock of the unfamiliar, whether caused by the need to speak foreign languages or to understand specific elements of a host culture (e.g., cuisine and manners). In the early 21st century, however, some experienced tourists increasingly are seeking out challenges that early travelers would have strived to avoid. This is particularly true of physical challenges that may generate discomfort (e.g., rudimentary means of travel such as skis, sleeping and living outdoors, and eating dried food.). To some extent, this phenomenon is a typical response from individuals who live in the current "risk society" (Beck, 1986, f.t. 2008), which strives to be risk-free. If so, the popularity of adventure travel among the general public could be partly rooted in the negative effects of the industrial and postindustrial lifestyle on human metabolism and the human brain.

Western societies have indeed erected a safety net of rules that also limit the potential for personal growth through initiative, risk, and creativity. This is compounded by a proliferation of social constraints that include incentives for self-censorship, intolerance, and conservatism. In a diverse society in which identity construction is both a matter of personal choice and juxtaposition, and is not bound to a single acceptable model, alienation is a latent enemy. To remedy this, the recreational cure promises to refocus an individual by replacing social constraints with stimuli from immersion in nature. Touted since the advent of the romantic movement, the beneficial effects of contact with nature have been amply demonstrated: reduced heart rate, lowered blood pressure, release of stress hormones, increased cognitive function (performance and creativity), and muscle relaxation (Kellert, 1993).

Outdoor activities, whose popularity has grown exponentially since World War II, have proved successful in every social stratum, primarily among the middle class, which has the time and can afford the necessary equipment. Outings create a thirst in participants for the unknown and risk that constitute adventure. Despite extensive literature on the concept of adventure, it remains poorly defined, often lumped in with other genres such as outdoor pursuits, ecotourism, and even expeditions (Buckley, 1997). Without a clear concept, researchers have strived to determine its parameters.

Some associate adventure with physical and mental challenges (France, 1997, p. 16; Mortlock, 2000, pp. 19–34; Swarbrooke, Beard, Leckie, & Pomfret, 2003;

Figure 7.2 Loss of reference points is a key component of adventure. (Photo by Alain A. Grenier.)

Weber, 2001: 360), while others build their analysis around risk (Barton, 2007; Goeldner and Ritchie, 2000, p. 721; Mortlock, 2000; Swarbrooke et al., 2003) and danger (Swarbrooke et al., 2003) or the perception of risk (Goeldner & Ritchie, 2000, p. 721). Whether subjective (i.e., a danger under the participant's control, such as the choice of equipment or co-adventurers) or objective (i.e., a risk beyond human control, such as the elements), risk is subject to the perception of the individuals involved, based on their experience, skills, reflexes, and physical and psychological endurance (Mortlock, 2000, p. 32).

Hunt (1989) maintains that adventure can be defined through a combination of various factors, including remoteness of the site, level of skill required, amount of exertion needed, and presence or absence of personal responsibility. Remoteness presumes the loss, to some extent, of familiar reference points, and thus safety. This loss generates circumstances amenable to discovery, in a tense ambiance, since the participant has no guarantees of the outcome (see Figures 7.2 and 7.3). Individuals who expose themselves to danger awaken all their senses that may have been numbed by daily routine. Unlike conventional tourism, characterized by the purchase of a predetermined experience, adventure tourism requires participants to contribute their own effort. Hence, the definition of adventure tourism is centered primarily on recreational activities that demand physical exertion and contact with nature (Weber, 2001, p. 360). Moreover, engaging in an unknown experience and outcome is a deliberate choice. This uncertainty over success of the activity is especially rewarding because it entails preserving a participant's physical and mental integrity.

Unlike conventional travel, adventure travel provides tourists with the opportunity for intense sensory stimulation (Muller & Cleaver, 2000, p. 156). This travel experience encourages individuals to achieve personal growth by using their abilities and skills. Where the conventional tourism industry intervenes to reduce and control negative effects on travelers, adventurers deliberately choose to limit intervention by the travel operator. The resulting experience obviously is built around varying degrees of risk.

Mortlock (2000, pp. 19–24) divides all types of adventure into four categories, from play (risk-free) to misadventure (negative risk). An adventure experience is positive when it results in noticeable gains, especially various sensations of well-being that may even extend to intoxication ["nature orgasm" (Viken, 1995, p. 81)]. An adventure experience is negative when the benefits are perceived as less than the effort extended. Ultimately, a negative experience may lead to very serious injury or even death. We find a logical progression, therefore, from a tourist outing (light adventure) to adventure per se (with a risk of failure). The travel industry's co-opting of the expedition concept to stimulate tourists therefore inevitably draws in risk.

Figure 7.3 Bringing tourists close to danger through adventure awakens all the senses. (Photo by Alain A. Grenier.)

7.3. From Adventure to Expedition

The adventure tourism industry's growing use in recent years of the word "expedition" raises a few questions about linking the concepts of "expedition" and "tourism." The etymological roots of the word *expedition* (from the Latin *expeditionem, de expedire*) trace back to the verb "to get rid of" and refer more to sending (parcels, for example) than to its common meaning today. The leading dictionaries of English (Oxford, Cambridge, Merriam-Webster) all concur in defining "expedition" as "a journey or excursion undertaken for a specific purpose" (Merriam-Webster, 2013). The meaning is subtler in French.

In 1694, the dictionary of the Académie française (cited in ARTLF, N.D.) first links "expedition" to a military action. In this same perspective, the Académie française (cited in ARTLF, N.D.) added in 1835 that this is [translation] "an enterprise of war requiring travel over a distance of varying length." The term "expedition" was used at the time primarily in reference to sea journeys. In the 20th century, an expedition also consisted of a journey [translation] "undertaken for a scientific, commercial or industrial purpose" (DAF, 1932–1935, cited in ARTLF, N.D.). In addition to the concepts mentioned above, especially that of a military operation, CNTRL (2008) adds that this may consist of a [translation] "hostile undertaking against someone or something" or a [translation] "scientific or tourist journey or outing." On this last point, Larousse dictionary (2008) specifies that an expedition is a [translation] "scientific voyage to a remote or inaccessible country, or a fairly significant and eventful tourist journey." Finally, in the familiar sense, it is defined as a [translation] "trip considered difficult or to a remote place."

A few concepts emerge from these definitions. The military aspect implies the participation of "troops" and their "movement" (CNTRL, 2008), which assumes group management. There are also concepts of distance—an expedition takes participants to distant or foreign and unknown lands, and this takes time; there is no quick return. The only concept missing from the definitions we have seen is that of enjoyment, a trip taken for pleasure. From a historical perspective, an expedition to some extent resembles "organized" trips by small groups of people, such as scouts (travel actually constituting migration), joined in a shared project that requires travel over great distances, with no guarantee of either return or success.

Between the earliest human journeys and the business and tourist adventures we know today, expeditions have evolved through various stages, each marked by a transformation in the "genetics" or nature of the undertaking. Table 7.1 maps the various groups that can claim to be on an expedition. In this illustration, the distinction among these various types of travelers—explorers (prehistoric and from the Age of Exploration), conquerors, professional adventurers,

Table 7.1. Changes over Time in the Primary Characteristics of Expeditions Based on the Various Types of Explorers

	Travelers				Tourists
	Explorers				
	Prehistoric	Age of Exploration	Conquerors	Professional Adventurers	Adventure Tourists
Primary motivations	• Survival (location/ discovery of new resources)	• Discoveries • Economic, political, and scientific	• Capture of resources	• Achieve records • Social distinction • Discoveries	• Personal distinction (social capital • Discovery
Stakes	• Domestic unit (family, village) • National	• Economic and political superiority	• Political power of the State (or a sovereign)	• Awareness of a cause	• Personal challenge
External aid (during the adventure)	• None	• None	• None	• Limited (in some phases)	• Almost continuous
Technology and logistical organization	• Simple • Functional autonomy	• Developed • Functional autonomy	• More developed • Functional autonomy	• Highly developed • Assistance	• Highly developed • Guide
Financing	• Dependent on local natural resources	• By private companies; the State (the society)	• By the State (the society)	• Sponsors (private)	• Self-financed
Risks	• Very high	• Very high	• Very high	• Average to very high	• Low
Benefits	• Group survival/emancipation		• National and personal merit	• Personal and national merit • Income	• Personal merit
		Antiquity	Modern day →		

and modern tourists—represents a fundamental, genetic shift in expeditions, due to the prime motivation, the stakes, the availability of outside aid or assistance, the type of technology and logistics used, financing, risks, and, finally, the benefits.

Three main findings emerge from this suggested reading of historical changes in the primary characteristics of expeditions, based on the main types of explorers:

1. A fundamental shift in the reasons for an expedition (from survival to political, social and personal emotions), which required:
2. A transformation in the nature of the activity, from a community action to a political action before being "commodified" into a consumer product. This presumes:
3. Benefits of a very different nature for those involved—the merit of personal and community survival as opposed to acquisition of political, and eventually social, capital. In this last case, only the individual benefits.

As we move through time (from left to right in Table 7.1), risk steadily declines but, paradoxically, increases in value disproportionately. Of all the changes observed in the "genetics" of expeditions, the most significant in relation to this study is personalization of the undertaking. While expeditions initially were group and social quests for collective survival and emancipation, they have now become individual and private ventures. The merit for participants appears to be associated primarily with the extreme nature of the environments transited, which evokes images associated with traditional expeditions. Those undertaking such a trip gain social merit primarily from this unstated historical association: The word "expedition" carries a collective cultural and historical mythology.

Today, the social boundaries that once separated mountain climbers, for example, from tourists are fading (Beedie & Hudson, 2003, p. 625) (see Figure 7.4). This is the crux of the barely concealed tension between professional explorers and tourists. That tension results from the overlapping between the interests of professional adventurers, who demand a measure of exclusive entitlement to the merit of their expeditions conducted at risk of life and limb (not to mention financial security), and those of adventure tourists, who are often able to engage in expeditions and extreme adventures solely by dint of their purchasing power. The nature of expedition travel can also vary. With no place left to plant a flag, some carry the torch of a cause as they strive to set a record (e.g., endurance, distance, originality) to raise public awareness of a cause. We are also witnessing the birth of a new class of adventurers: scientific tourists or "scientourists," who organize pleasure travel around the opportunity to acquire professional knowledge.

Figure 7.4 Treading lightly with a free heart, today's hikers draw part of their enjoyment in this activity from its association with the mythology of explorers. (Photo by Alain A. Grenier.)

This blurring of genres occurs because [translation] "technological progress has made possible performance previously considered unattainable," notes Perrin (2002, p. 70). For professional adventurers to differentiate themselves from adventure tourists, they require ever more daring: [translation] "venturing forth without a safety net" (Perrin, 2002, pp. 70–71). Taking the example of Himalayan travelers, Raspaud (2002, p. 101) notes that [translation] "the line between the elite [. . .] and everyone else, especially those on commercial climbs, involves [. . .] the refusal to use artificial oxygen supplies and thus a step of sorts from away from technology, in addition, of course, to the practice of seeking new ascents."

However, adventure tourism does not generate only inconvenient professional explorers. Since not all professional adventurers can make a living from their expeditions (Le Scanff, 2000, p. 38), many compromise and work for operators as expert guides. Such an alliance offers several advantages. As shown by Figure 7.5, these adventurers can capitalize on the media attention generated

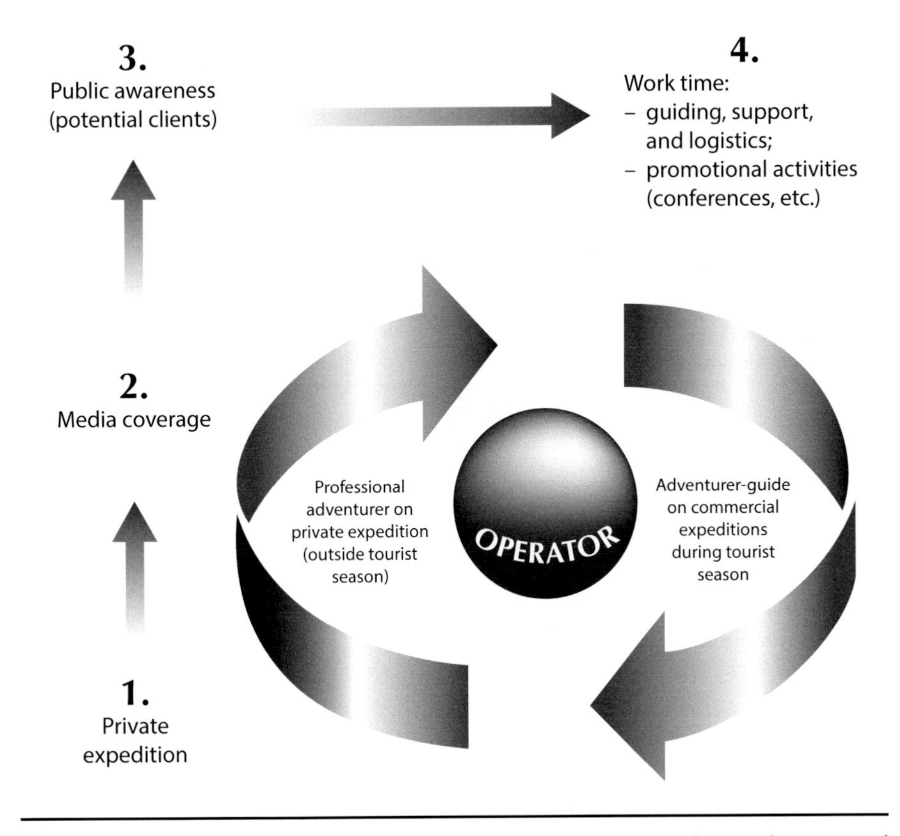

Figure 7.5 The business link between adventure tourism and expedition travel (professional adventurers).

Table 7.2. Survey of Values and Factors Sought in Expedition Travel, According to Various Authors

	Values/Factors Sought in Expedition Travel	1	2	3	4	5	6	7
Physical	Physical activity		X					X
	Personal transcendence (drawing on physical skills not used in daily life)	X	X		X	X		
	Outperforming others (competition), need to dominate	X	X	X				
	Drive for independence							X
	Appetite for risk							X
Mental	Quest for pleasure	X				X		X
	Need to bolster personal self-esteem, self-confidence	X			X	X		
	Need for to take stock, to clean house (search for silence/solitude)	X		X				
	Need to feel alive/face one's mortality	X			X	X	X	
	Quest for internal peace (through contact with the natural environment)	X			X	X		
	Develop one's own life philosophy (find meaning in one's life)	X					X	
	Replace the ordinary with the extraordinary		X				X	X
Identity	Acquire experience/skills							X
	Affirm one's difference	X						
	Affirm one's combativeness					X		
	Replace the ordinary with the extraordinary		X				X	
	Gender affirmation (masculinity for men) Character affirmation (for women)	X					X	
	Social recognition (acquire a reputation, distinction, status)	X				X		
	Adventure as a deliberate choice					X		

1. Le Scanff (2000); 2. Le Breton (2002); 3. Perret (2002); 5. Piccard (2002); 6. Mortlock (2000) 7. Wiger (2002).

by their personal expeditions to attract clients and work when not exploring or breaking records. Operators benefit from the media spotlight on their celebrity guides. In turn, these professional adventurers benefit directly from contact with a target audience with whom they can share their experience. They also gain financial support they need for their own personal expeditions they lead between commercial "tourism adventures." This type of association can, however, create discord with purist adventurers who lament and occasionally condemn the commercial aspect of expedition travel.

Professional and tourist expeditions also differ by the types of relationships they foster among participants. In professional expeditions, individuals form interdependent bonds based on each person's complementary skills, expertise, and experience. Responsibility for the group's psychological and physical health rests with each participant. Everyone is interconnected through horizontal links of equality and responsibility. In relation to the collective body they form, the individuals involved constitute the members of the expedition.

In a commercial expedition, the group also consists of individuals brought together by a shared geographic goal. However, participants may also be motivated by differing personal goals. Although attainment of the goal depends in part on each person's efforts and skills, as in a professional expedition, it also relies on experts hired to guide the group. There is a hierarchy (i.e., vertical links) and a transfer of personal responsibility, even if only partial, from tourists to guides. The initiative consists primarily of participants who join it, and its success depends on paid expertise.

Since they are driven by the need to be different, adventure tourists may seek novelty in an extreme adventure (i.e., an expedition), where professional adventurers look out for their safety. Extreme adventure blends the two genres.

7.4. The Role of Expeditions in Modern Society

In an attempt to understand the profile of needs and motives driving adventure and expedition travel, we identified seven key essays drawn from the literature in French and English. Some are based on interviews with professional adventurers or guides in training. Others present an analysis based on the literature. Using the content analysis method, we discovered 14 motives. Our findings are summarized in Table 7.2. For clarity's sake, these motives are divided into three categories: physical, mental, and identity. We recognize the interrelation of these elements but still maintain the distinction to facilitate comprehension.

7.4.1. Physical Motivation

In this area, we find that adventurers need to connect fully with their physical being. The additional effort required by the extreme aspect of the experience makes people fully aware of their body, through physical hardship (e.g., muscle pain, hunger, thirst, and extreme temperature), "We must feel to exist," says Piccard (2002, p. 15). [translation] "Physical pain is [. . .] one way to experience the body, to feel alive," explains Le Scanff (2000, p. 62), who also rejects Western society's intolerance for pain. In this way as well, extreme adventure not only contradicts certain aspects of Western society but also restores a certain balance—in this case, between the body, which is usually overprotected from pain, and the need to accept this as a condition of life.

Engaging in these activities is also often portrayed as compensating for living in an orderly, calm society, perceived as sterile and bland (Le Breton, 2002, p. 75), and provides an opportunity for affirming one's gender identity. Like sports in general, adventure tourism and especially expedition travel are venues for some men to affirm masculinity. (Without exception, male sports capture a greater share of media attention. This reality is also transferred to the phenomenon of adventure tourism, where participation forms part of two different gender identities.)

Challenging terrain and climate—and sometimes culture or the sociopolitical context—allows participants to take on the idealized roles of "explorer" and "pioneer" (Gyimóthy and Mykletun, 2004, p. 865). In a study of travel accounts by modern adventurers, Gyimóthy and Mykletun (2004, p. 865) note that these men seek to identify themselves with heroic characters from history: loners with bruised, wind-chapped, sunburnt faces. The already highly masculine connotation of sports is carried even further in high-risk activities (Le Scanff, 2000, pp. 32–33). I have witnessed male tourists bare their heads (and sometimes more) just long enough to accumulate a coating of frost, for a photograph that strongly conveys their endurance. Not shaving while on expedition is another way of reinforcing images of virility. Similarly, [translation] "many mountain climbers (especially men . . .) conceal the initial signs [of altitude sickness], unwilling to appear 'weak'," notes Wiget (2002, p. 41). Some women instead may seek strength of character through contact with adventure and male competitors.

This implies that expedition travel, as an extreme activity, constantly forces people to face not only themselves (i.e., the mental aspect, which we will see shortly) but also their community. Le Breton (2002, p. 77) points out that the logic of adventure [translation] "is the *individual challenge,* a personal quest for legitimacy, notoriety, recognition that co-opts [nature] as an element of danger to be overcome by making oneself as vulnerable as possible while believing one still has 'the right stuff.'" Tension therefore exists between the adventurer and nature. Competition with other adventurers comes second, and its importance depends on the relationship with public recognition and social status. This factor is also present in tourism as reflected in the choice of destination and type of activities.

7.4.2. Mental Motivation

Although expedition travel is experienced primarily as a physical challenge, it requires great mental strength as well. Expedition travel presents greater opportunity for rewarding physical endurance than proving one's psychological strength—people are usually more interested in testing their physical abilities than their psychological endurance. Isolation combined with dependence on the participants' own capabilities may lead them to question their own existence. [translation] "In our society increasingly devoid of emotion and personal responsibility, just enough stimulation remains available to nurture routine, but not *Awareness*," writes Piccard (2002, p. 15; author's emphasis). Expedition travel therefore is not so much an opportunity to test one's mental strength as to provide the mind with the "elbow room" it lacks in daily life. [translation] "Solitude exerts a pull, an almost mandatory transition, a trial that helps us more fully discover our own limits and find solutions we would never discover if we traveled with someone else," explains Le Scanff (2000, pp. 77–78). The void created in extreme wilderness by the absence of others (society) allows a person to refocus, to find his center again. We see the same phenomenon in nature tourists who, following immersion in the natural environment, often make philosophical observations in a more pensive state of mind.

7.4.3. Identity Motivation

Finally, physical and mental motivation is instrumental in constructing self-identity and in gaining recognition of others (Berger & Greenspan, 2008, p. 92). The analysis conducted by Heuzé, Le Scanff, Zimmermann, Rosnet, & Vion (2002, p. 50) on the motives of 13 mountain guides and 58 trainees in a risk management study found that the personality of a mountain guide is characterized by high anxiety, low self-esteem, and poor self-affirmation. In addition, guides in training had a greater need for social recognition (Heuzé et al., 2002, p. 50).

When interviewed by Le Scanff (2000, p. 29), accomplished French polar explorer Jean-Louis Étienne attributed his own ambitions as an adventurer to [translation] "a desire to prove to himself and others that he was as capable as anyone else," and to loss of self-esteem triggered by certain disappointments (Ibid.). Étienne (in Le Scanff, 2000, pp. 44–45) also believes that extreme adventure arises in particular from overambition driven by low self-esteem:

> [translation] Those who sail around the world alone are also seeking self-affirmation through their own vessel, but they are entered in an official race. Polar adventurers, on the other hand, have a much greater need that they seek to meet through their mission. These people are acting out their quest for heroism, putting themselves front and center in a display of much grander unquenched ambition and ego.

Le Breton (2002, p. 74) carries this thinking further:

> [translation] . . . our societies formed of individuals require everyone to prove the legitimacy of his or her existence in a competitive world striving for efficiency that places little value on a quiet stroll and the peaceful pleasure of just being. The personal passions that now cast high-risk activities in this surprising social landscape arise from circumstances that often place great importance on proving to oneself the value of one's own existence. As the producers of their own identity, individuals are dependent on outside approval as reassurance of their own value.

In play, individuals enter a different reality defined by their own rules, values, and expectations (Gyimóthy & Mykletun, 2004, p. 859). Jafari (1988) talks of the suspension phase, the time during which tourists are separated from their routine environment and thus their responsibilities, and enter into a state of suspension (i.e., play). They can drop the mask of their everyday life and show how they would like to be seen: an adventurer, in our case here, a person motivated by challenge, daring, bravery, and tenacity, as reflected in personal travel choices. This personal revelation emerges from the trial, the temporary break from the familiar (Le Breton, 2002, p. 73). As a result, tourists tend to take on greater risk, but this is perceived rather than real risk.

In Western societies, which tend to depersonalize and sterilize the soul, expedition travel therefore provides a means of producing small and large models for acquiring or enhancing personal identity in large leaps. At the very least, people believe that an expedition helps them carve out their place in the sun.

Studies of expedition tourists are needed to show whether they share with professional adventurers the values and factors reported in Table 7.2. There is reason to believe that both groups have similar motivations but differ in the way they meet these needs. The price tourists pay for so-called expedition packages is nothing less than the cost of admission. The skill and merit associated with the experience are non-negotiable.

There comes in an expedition a point where, based on involvement and risks, one ceases to be a tourist and instead becomes an expedition traveler in the traditional sense—a conqueror who has dominated a specific extreme geographic area (e.g., mountain summit, or ocean) despite the risk to physical, psychological, and social integrity. When we consider the horizontal and vertical links that differentiate the nature of teams, we find we not only have to review the concept of expedition but also, perhaps more important, we have to review the aspects, previously discussed in this chapter of men and women who take part in expeditions. In this sense, mountain climbing, for example, already provides solutions by differentiating guides (i.e., the professionals) from tourists (i.e., the clients).

7.5. Discussion

From amusement parks to reality TV, adventure, once limited to a handful of individuals and extreme geographic environments, is now present everywhere. It even touts its management models for business (see Perkins, Holman, Kessler, & McCarthy, 2003). Adventure, to some extent, has become the new benchmark, affecting all layers of society (Le Scanff, 2000, p. 9). Through the risks it entails and the decision making it demands, adventure travel impels people to use all their senses and draw on all their abilities. As people dangle at the end of a line, which was Simpson's case (1988), adventure forces them to live—to face themselves, to take control of their lives. Where traditional society was built on participation by each member, Western society has become overorganized to eliminate risk and thereby wipes out the very vital space in which humans might still feel alive.

When members of modern Western society can no longer engage their full potential in the routine, workaday world, they can turn to the playground provided by adventure tourism [the suspension moment, as in Jafari's model (1988)] to access enriching, stimulating experiences. Adventure tourism pushes up the level of risk to where fun begins, whether it be in an amusement park, exploring a red-light district, or scaling a mountain. The purpose of adventure is often to push oneself to the limit (Gyimóthy and Mykletun, 2004, p. 855), which differs among individuals and the challenges they take on. Adventure is a relative concept dependent on each person's abilities, which explains its appropriation by the tourism industry.

Tourism is a special phenomenon. In a constant quest for distinctiveness, it reinvents itself in comparison with other sectors, which it imitates. In combination with the sports world, it offers unusual but supervised activities that provide the necessary safety and logistics for safely testing oneself. Access to remote natural environments encourages the creation of a more active vacation model that combines physical activity and sports with travel to achieve distinctiveness. This combination is made possible by:

1. Greater environmental awareness that has grown since the 1980s
2. Greater value placed on a more active approach to travel and recreation that combines physical activity and sports with travel to achieve distinctiveness
3. Democratization of sports technology, which simultaneously leads to fragmentation of the sports sector

Technology now makes adventure travel more accessible to tourists, both physically (i.e., venturing into remote and harsh places is now possible) and socially (i.e., society values risk takers). High-performance and extreme sports can now be down-sized and divided into more digestible bites. Mountain climbing has given rise to wall climbing and mountain hiking or trekking. Other modes of travel such as cycling or watercraft (e.g., kayak or canoe) also fit the adventure/expedition approach when they occur in difficult terrain, reducing the outside support required and extending the length of the activity.

Mountain, marine, polar, and desert regions provide perfect playgrounds for experiencing risk and unexpected adventure because they provide difficult terrain that is a challenge to access and relatively unexplored and is marked by extreme climate and dangerous wildlife. To stand out from other tourism products, adventure travel operators are quick to use the words "adventure" and "expedition," often incorporating them into their business names as a mark of distinction. Among polar travel operators, we found many examples: Antarpply *Expeditions,* Aurora *Expeditions, Adventure* Associates, *Adventure* Network International/Antarctic Logistics & *Expeditions, Xplore Expeditions,* G.A.P. *Adventures,* Lindblad *Expeditions,* Ocean *Expeditions,* Oceanwide *Expeditions,* Orion *Expedition* Cruises, Pelagic *Expeditions* LTD, Polar Star *Expeditions,* Quark *Expeditions,* Sea & Ice & Mountains *Expeditions,* Spirit of Sydney *Expeditions* Pty Ltd, Waterproof *Expeditions,* Zegrahm *Expeditions.*

Other operators emphasize the concepts of exploration and discovery: Hanse *Explorer* Gmbh & Co. KG, Heritage *Expeditions,* and Voyages of *Discovery.* Finally, a few operators use extreme geographic references: *High* Latitudes Limited, Latitude Océane. Some also appeal to *playful adventurers,* those tourists who differ from the masses by the extreme daring of their choices, even if those choices involve very close supervision.

Price (1978, cited in Beedie and Hudson, 2003, p. 627) believes, however, that adventures cannot be planned. Yet this is not how adventure tourism is presented and sold to the public. As Beedie and Hudson (2003, p. 627) point out, the more tightly detailed and planned a trip is to facilitate logistics, the further the planned experience strays from the concept of adventure. Beedie and Hudson (2003, p. 627) identify three factors that can transform and "commodify" adventure into a tourism product, packaged, marketed, sold, and purchased to be consumed:

1. Handing over control to experts
2. Proliferation of promotional material (e.g., brochures)
3. Application of technology in the adventure

They maintain that these factors create a comfort zone between the normal daily urban domestic environment and the extraordinary experience of adventure travel (Beedie & Hudson, 2003, p. 627). From this emerges the idea that applying the word "expedition" to adventure tourism is sacrilege: "Tourism" strips away the sacred merits of grand adventure.

On the other hand, adventure tourism implies what Berger and Greenspan (2008, p. 91) call the *intellectualization* of travel—a reminder of sorts of the Grand Tour in which the travel is justified by its educational aspect and *professionalization* (Munt, 1994; Stebbins, 1982). This type of experience implies the presence of professional leaders whose credibility, experience, and *curriculum vitae* establish and justify their position and responsibilities (Berger & Greenspan, 2008, p. 92). These beliefs are sold as a guarantee of safety for clients and at the same time create a sacred aura around the expert guides and the product (as explained by MacCannell, 1999). These expert guides therefore should not concern themselves with the vagueness surrounding the concept that, until recently, was theirs alone.

To a great extent, adventure tourism, whether it is or is not part of a consumption process (i.e., buying a package trip, as opposed to organizing one's own expedition), contributes to the identity-creation process for tourists just as it did in the past for explorers, conquerors, and, more recently, professional adventurers. Modern society provides a very fluid model of identity definition (McCracken, 1986). Social players have the freedom and the ability to create and re-create identities of their choosing, almost instantly (Berger & Greenspan, 2008, p. 90). Buying an adventure travel package associated, even in name only, with expeditions contributes to this identity creation. Professional and novice guides still maintain their distinctiveness through their approach to adventure; purchasing a place on a trip is no substitute for paying one's dues over time.

The tension that divides professional adventurers over the arrival of tourists on their sacred ground, as justified as it may be, therefore is vain. It does not challenge the merits of either camp, except perhaps in the eyes of a distant spectator. Quite the contrary, expedition tourism adds merit to the guides, whose responsibility is no longer limited to just their own life and the lives of their companions, but now extends to those of novice clients, whose lack of experience poses additional risks to success of the undertaking.

7.6. Conclusion

In questioning the legitimacy of using the term *expedition* for marketing purposes, at issue is the very essence of the expedition experience and the merit of the participants, whether professional or tourist. Can travel by an organized group of tourists, even with a professional adventurer as guide, constitute an expedition? While the answer remains largely a matter of opinion, use of the word *expedition* by the tourism industry provides open access to the images of the mythical world of exploration. Through these "borrowed" images and illusions is tourism built. Whether of the expedition or conventional variety, tourism is an exploratory experience that distances travelers from their usual reference points and inevitably leads to immersion in discovery through travel.

Western societies, obsessed with safety and security in daily life, are driving some people to seek new experiences involving a degree of risk. Adventure tourism meets this need in part. The distinction required in tourism, combined with the democratization in the use of technology and accessibility to natural environments, should lead one day to even more ambitious and extreme adventure trips: expedition tourist packages.

Association of the word *expedition* with the tourism sector does indeed create tension with the professional expedition sector. We have also suggested that the tension existing between certain professional adventurers and adventure tourists may arise from the modification and adaptation of the concept of expedition, which is related to the current trend of seeking more sports-oriented and extreme recreation. This leads to resistance by professional adventurers who fear the trivialization of their own extreme experiences and the merit gained from them. When the tidal wave of mass tourism washes in, who pays attention to extreme adventurers?

Yet mass-market tourists cannot follow in the steps of professional adventurers, since the very concept of tourism implies a structured experience with a planned start and end. Expedition travel requires horizontal links of equality and interdependence among members, whereas adventure tourism, even when it borrows certain aspects from expedition travel, cannot eliminate the vertical links because it needs a hierarchical structure.

What operators call an expedition does, however, give individuals the opportunity to leave behind the crushing banality of daily life, and it triggers in them strong emotions that awaken all their senses, skills, and physical, mental, and emotional abilities during a specific window of recreational time that is tourist travel. Participants come out of this experience alert, as if they had left the numbed-out, bitter daily world, at least for a time.

Any living concept takes on the hues of its era. The democratization of society as well as its technology and practices at the dawn of this new millennium is opening a whole new range of human possibilities. In a world where everything becomes possible, expedition travel is no longer defined by the objectives to be attained but by the burden of the logistical organization of the venture and the time required to carry it out—and that sets it apart from conventional tourism.

If the sinking of the *Explorer* in Antarctica had resulted in loss of human life, there is little doubt the accident would have discouraged future tourists, as occurred in the years following the crash of Air New Zealand flight 901. By reminding us that adventure entails risks, without causing the death of participants, the sinking of the *Explorer* instead raised the profile of adventure or expedition tourism. While adventure conducted in a commercial framework may somehow dilute the attention once focused on explorers, it does nothing to diminish their merits. It simply forces intrepid travelers, as well as adventure expedition enthusiasts, to rethink and redefine the boundaries of their experiences and stories.

References

Acott, T. G., La Trobe, H. L., & Howard S. H. (1998). An evaluation of deep ecotourism and shallow ecotourism. *Journal of Sustainable Tourism, 6*(3), 238–253.

ARTLF (N.D.). Dictionnaires d'autrefois—Dictionnaire de l'Académie française, 6th ed. (1835). ARTLF Project, University of Chicago. http://artflsrv02.uchicago.edu/cgi-bin/dicos/pubdico1look.pl?strippedhw=exp%C3%A9dition

Barton, B. (2007). *Safety, risk & adventure.* London: Paul Chapman & SAGE.

Basset, C. A. (2009). *Galapagos at the crossroads: Pirates, biologists, tourists, and creationists battle for Darwin's cradle of evolution.* Washington, DC: National Geographic.

Beck, U. (1986; French translation 2008). *La Société du Risque—Sur la voie d'une autre modernité.* Paris: Flammarion.

Beedie, P., & Hudson, S. (2003). Emergence of mountain-based adventure tourism. *Annals of Tourism Research, 30*(3), 625–643.

Berger, I. E., & Greenspan, I. (2008). High (on) technology: Producing tourist identities through tehnologized adventure. *Journal of Sport & Tourism, 13*(2), 89–114.

Bourdieu, P. (1979). *La distinction—Critique social du jugement.* Paris: Les Éditions de Minuit.

Boyer, M. (1995). L'invention de distinction, moteur du tourisme? Hier et aujourd'hui. *Téoros, 14*(2), 45–47.

Buckley, R. (1997). *Adventure tourism.* Royaume-Uni: CABI Publishing.

Burton, F. (1998). Can ecotourism objectives be achieved?" *Annals of Tourism Research, 25*(3), 755–757.

Butcher, J. (2005). The moral authority of ecotourism: A critique. *Current Issues in Tourism, 8*(2),114–124.

Clifton, J., & Benson A. (2006). Planning for sustainable ecotourism: The case for research ecotourism in developing country destinations. *Journal of Sustainable Tourism, 14*(3), 238–254.

CNRTL—Centre national de ressources textuelles et lexicales (2008). Expedition. Retrieved 10.07.2009 from www.cnrtl.fr/definition/expedition.

DAF—Dictionnaire de l'Académie française, 8th ed. (1932–1935). Quoted in ARTLF (N.D.). Dictionnaires d'autrefois—Dictionnaire de l'Académie française, 6th ed. (1835). ARTLF Project, University of Chicago. Retrieved 10.07.2009 from http://artflx.uchicago.edu/cgi-bin/dicos/pubdico1look.pl?strippedhw=expedition

Diamantis, D. (1999). The concept of ecotourism: Evolution and trend. *Current Issues in Tourism, 2*(2), 93–122.

D'Orso, M. (2002). *Plundering paradise: The hand of man on the Galapagos Islands.* New York: Harper Collins.

France, L. (1997). *Sustainable tourism.* London: Earthscan.

Goeldner, C. R., & Brent, R. (2000). *Tourism: Principles, practices, philosophies.* Toronto: John Wiley.

Grenier, A. A. (2009). Conceptualisation du tourism polaire: Cartographier une experience aux confins de l'imaginaire. *Téoros, 28*(1), 7–19.

Grenier, A. A. (2004). *Nature of nature tourism* (Dissertation). Acta Universitatis Lapponiensis, 72. Faculty of Social Sciences, University of Lapland, Rovaniemi, Finland.

Grenier, A. A. (1998). *Ship-based polar tourism in the Northeast Passage.* Rovaniemi, Finland: University of Lapland, Publications in the Social Sciences.

Gyimóthy, S. and Mykltun, R. J. (2004). Play in adventure tourism: The case of Arctic trekking. *Annals of Tourism Research, 31*(4), 855–878.

Heuzé, J.-P., Le Scanff, C., Zimmermann, S., Rosnet, E., & Vion, J.-P. (2002). Étude interactionniste de la gestion des risques dans des activités professionnelles de montagne. In M. Braddeley (Ed.), *Sports extrêmes—Sports de l'extrême* (pp. 43–55). Genève: Académie internationale des sciences et techniques du sport, Georg Éditeur.

Jafari, J. (1988). Le système du touriste: Modèles socio-culturels en vue d'applications pratiques et théoriques. *Loisir et Société, 11*(1), 59–80. Montréal: Presses de l'Université du Québec.

Hunt, L. (1989). *In search of adventure: A study of opportunities for adventure and challenge for young people.* Guildford: Talbot Adair Press.

Kellert, S. R. (1993). Experiencing nature: Affective, cognitive, and evaluative development in children. In S. R. Kellert; E. O. Wilson, and S. McVay (Eds.), *The biophilia hypothesis* (pp. 117–152). Chicago: Island Press.

Larousse (2008). Expédition, Retrieved 11.03.2013 from Dictionnaire de français Larousse, http://www.larousse.fr/dictionnaires/francais/exp%C3%A9dition/32232.

Le Scanff, C. (2000). *Les aventuriers de l'extrême.* France: Calmann-Lévy.

Le Breton, D. (2002). Ceux qui vont en mer: Le risque et la mer. In M. Braddeley (Ed.), *Sports extrêmes—Sports de l'extrême* (pp. 73–82). Genève: Académie internationale des sciences et techniques du sport, Georg Editeur..

Lee, T.-H., & Crompton, J. (1992). Measuring novelty seeking in tourism. *Annals of Tourism Research, 19*(4), 732–751.

MacCannell, D. (1999). *The tourist—A new theory of the leisure class.* Berkeley: University of California Press.

McCracken, G. (1986). Culture and consumption: A theoretical account of the structure and movement of the cultural meaning of consumer goods. *Journal of Consumer Research,* 71–84.

Merriam-Webster (2013). Expedition. Retrieved 11.03.2013 from Merriam-Webster Online, http://www.merriam-webster.com/dictionary/expedition

Mortlock, C. (2000). *The adventure alternative.* United Kingdom: Cicerone Press.

Muller, T. E., and Cleaver, M. (2000). Targetting the CANZUS baby boomer explorer and adventurer segments. *Journal of Vacation Marketing, 6*(2), 154–169.

Munt, I. (1994). The "other" postmodern tourism. Culture, travel and the new middle class. *Theory, Culture and Society, 11*, 101–123.

Perkins, D. N. T., Holman, M. P., Kessler, P. R., & McCarthy, C. (2003). *Leading at the edge—Leadership lessons from the extraordinary saga of Shackleton's Antarctic expedition*. New York: Amacom.

Perret, D. (2002). Sport ou regard « extrême »? In M. Braddeley (Ed.), *Sports extrêmes—Sports de l'extrême* (pp. 5–7). Genève: Académie internationale des sciences et techniques du sport, Georg Editeur.

Perrin, E. (2002). Rapport du Groupe de travail Société et technologie. In M. Braddeley (Ed.), *Sports extrêmes—Sports de l'extrême* (pp. 69–71). Genève: Académie internationale des sciences et techniques du sport, Georg Editeur.

Piccard, B. (2002). Le sport extrême: Une école de vie? In M. Braddeley (Ed.), *Sports extrêmes—Sports de l'extrême* (pp. 9–16). Genève: Académie internationale des sciences et techniques du sport, Georg Editeur.

Price, T. (1978). Adventure by numbers. In K. Wilson (Ed.), *The games climbers play* (pp. 646–651). London: Diadem. Cited in P. Beedie and S. Hudson (2003), Emergence of mountain-based adventure tourism, *Annals of Tourism Research, 30*(3), 625–643.

Raspaud, M. (2002). Himalayisme et usage de la technologie. M. Braddeley (Ed.), *Sports extrêmes—Sports de l'extrême* (pp. 93–101). Genève: Académie internationale des sciences et techniques du sport, Georg Editeur.

Simpson, J. (1988). *Touching the void*. United Kingdom: Jonathan Cape Ltd.

Stebbins, R. A. (1982). Serious leisure: A conceptual statement. *Pacific Sociological Review, 25*, 251–272.

Swarbrooke, J., Beard, C., Leckie, S., and Pomfret, G. (2003). *Adventure tourism: The new frontier*. Oxford, UK: Butterworth/Heinemann.

Weber, K. (2001). Outdoor adventure tourism—A review of research approaches. *Annals of Tourism Research, 28*(2), 360–377.

Wiger, U. (2002). L'altitude extrême: Un risque médical évident. M. Braddeley (Ed.), *Sports extrêmes—Sports de l'extrême* (pp. 39–42). Genève: Académie internationale des sciences et techniques du sport, Georg Editeur.

Viken, A. (1995). Tourism experiences in the Arctic—The Svalbard case. In M. Johnston and C. M. Hall, *Polar tourism: Tourism in the Arctic and Antarctic regions*. Toronto: John Wiley.

Part Two

Extreme Situations

Chapter 8

The Project Front End: Financial Guidance Based on Risk

Frédéric Gautier

What happens before a project effectively begins? What consequences are there for project performance and preparation within organizations undertaking projects? How is project preparation organized? Numerous studies emphasize the stakes of preliminary phases of a project. In the framework of the Twingo project, Midler (1993) presents "the battle of profitability" (p. 26), "first battle done by the project," and the way in which such battle was waged. At each stage of the project, dialogue between business divisions and specification of hierarchical criteria caused various battles concerning profitability. The significance of activities prior to detailed design and to new product development is explicitly clear (Cooper & Kleinschmidt, 1987) in the framework of the *NewProd* study of approximately 200 Canadian firms. According to Cooper and Kleinschmidt, project definition and upstream activities make up one of the three main factors of success for new products. Another study carried out by the same authors (Cooper & Kleinschmidt, 1996) on 161 North American businesses, which were in such varied industries as chemistry, equipment and machine tools, food processing, electronics, and automotive equipment, shows that the development process, and more particularly activities performed upstream of a project itself (e.g., business activities and decision-making stages and points), are the main factor of success in new-product development projects. With regard to project steering, the analyses of Fray, Giard, and Stokes (ECOSIP, 1993) highlight differentiated control systems before and in the process of a project. In automotive industry projects, the authors distinguish:

- A first phase whose objective is to define product specifications, industrialization pattern, and overall budget, and during which financial decisions essentially focus on constructing alternative scenarios based on technical, industrial, and financial stakes and risks
- A second phase, during which lock-in occurs because meeting schedule and costs are priority

All of these studies and statements lead to questions about activities prior to a project's implementation and how these activities are managed and governed. A time-related approach, however, leads to many activities concerning research techniques and processes and strategic reflection. Our discussion concerns more particularly the project front-end phase, its characteristics, and its methods for guidance.

8.1. The Project Front End: Learning Integration and Uncertainty

According to ECOSIP (1993, p. 147), the project front end is "a phase of elaboration of requirements that define product specifications, industrialization pattern and overall budget. . . . Financial planning essentially focuses on construction of cogent alternative scenarios by highlighting, in every case, technical, industrial and financial stakes (as well as risks incurred)." A significant characteristic of the project front end that differentiates it from the project proper is that it is either stop or go, that it ends with a decision whether or not to launch a project on the basis of a technical, industrial, and financial assessment.

8.1.1. An Organizational View of the Project Front End

Project management has been applied to the development of new products or services and used to reduce development time, which led us to investigate upstream project phases and, especially, the project front end. The notion of project front end, referred to as a fuzzy front end, appears in the analyses by Smith and Reinertsen (1991). This phase, prior to the development of a project, comprises three phases, as illustrated in Table 8.1.

Table 8.1. The Three Phases of the Project Front End

Phase 1	Prepare project proposal	Target consumers Target application Keystone advantages
Strategic filter		
Phase 2	Prepare business plan	Technical feasibility Marketing and economic feasibility Financial projections
Economic filter		
Phase 3	Prepare detailed project plans	Specifications Project budgeting Project scheduling

Smith and Reinertsen (1991) emphasize that research of new technologies, market research, and determining strategy are outside a project's front end. This amounts to defining the project front end as a phase of preparation and not of exploration. This is what the definition proposed by Gautier & Lenfle (2004, p. 17) emphasizes: The project front end relative to a project for the design and development of a new product is defined as "making of a new product or service development proposal including value hypotheses, technical and technological hypotheses, and hypotheses of industrial solutions." The essential characteristic of this definition is the decision to undertake or stop a project. Such an organizational definition of the project front end highlights the links between the project and its parent organization and the following organizational characteristics:

- A project front end implies a close cooperation with the parent organization, which provides the resources for the project front end.
- Work performed during the project front end depends greatly on a parent organization's objectives, as well as on whether the parent organization makes the decision to launch the project based on the project front end team's work; these objectives are likely to vary according to those of the permanent organization.
- Information is gradually acquired through preparative work processes.

This organizational approach (Andersen, 2008) distinguishes itself from the administrative, or task, perspective, which is used in certain analyses. The work of Khurana & Rosenthal (1997, 1998) falls within the administrative perspective because their approach attempts to formalize the project front end, relative to projects concerning design and development of new products, from activities to be performed. Though these analyses emphasize such important characteristics of the project front end as establishing a new multifunctional team deciding whether or not to launch a project, they are fuzzy about which tasks consist of exploration activities and which of preparation, and they render impossible the distinction between the project front end and other upstream project activities. In a more general perspective, the *PMBOK® Guide* (PMI, 2008) defines the process of project initialization in terms of activities: developing a project charter (aimed at authorizing the project and documenting initial requirements) and identifying stakeholders. However, this approach barely mentions the fact that the decision whether or not to launch a project is a real decision based on resource allocation and made by the permanent organization and that, accordingly, project initialization cannot be outside the parent organization.

On the basis of Bower's classical analysis (1970) of resource allocation processes within large organizations, the project front end appears to be a process consisting of a set of subprocesses managed at parent organization level, the decision whether or not to launch a project is inseparable from the strategic decision-making process of the parent organization, and multiple agents, at various levels of the parent organization or partner organizations, are involved in phases of the project preparative process. In this respect, Bower's work (1970) emphasizes that the project front end is a phase integrating multiple sources of organizational logic and knowledge leading to financial projections in terms of value and costs, technical and technological hypotheses, and hypotheses of industrial solutions. On the grounds of such hypotheses, the parent organization can make the decision whether to launch a project.

8.1.2. Integration of Knowledge in a Context of Uncertainty

There exists a fundamental difference between a project and the project's front end. A project is indeed characterized by focusing energies to meet a clear objective stated in terms of specifications, cost, and deadlines. This is not the case for a project front end, which seeks to determine relevant targets. A project front end aims at formulating a problem rather than solving it, and the way in which problems are solved during the course of a project closely depends on the way in which problems have been formulated. The whole scope of the organizational stake in the project front end is to mobilize a multifunctional team in order that the problem is defined in its multiple dimensions.

Moreover, a project front end is characterized by strong uncertainty, which creates specific difficulties. The project management literature focuses on risks and their management. However, the scope of managing uncertainty is broader: It encompasses potentially damaging consequences as well as potentially beneficial opportunities and does not consider only defensive management. Uncertainty needs to be analyzed from the point of view of the parent organization and of organizational services that make the decision regarding the launch of the project. According to Andersen (2008), uncertainty may result from various reasons:

- A lack of information
- A lack of knowledge, because all elements of the problem are not completely understood
- A lack of control that may be related to operational elements, tasks, or contextual and environmental elements.

Galbraith's studies (1973) show that uncertainty is a central variable of organizational design. Uncertainty is defined (Galbraith, 1973, p. 5) as "the difference between the amount of information required to perform the task and the amount of information already possessed by the organization." When available information is not sufficient, an organization opts for a strategy to decrease the quantity of necessary information (i.e., to create a *slack* resource or autonomous tasks) or to increase the information-processing capacity of the organization (i.e., invest in information systems or create lateral relationships in the form of a "task force"). In the framework of a project front end, those analyses enable consequences to be contemplated. Uncertainty cannot be managed unless a project front-end team understands its causes from the perspective of the parent organization. Uncertainty relative to a lack of information can be managed by implementing a multifunctional team that mobilizes a number of experts from the organization and includes stakeholders. More fundamentally, the project front end implies integration of knowledge made possible through compromise negotiated among various business divisions and agents to deal with the gaps in knowledge. In this regard, Iansati's work (1998) on integration can be considered project front-end theory. In the framework of projects on design and development of new products, Iansati (1998, p. 21) defines integration as "the set of investigation, evaluation and refinement activities aimed at creating a match between technological options and application context." Two categories of knowledge are distinguished:

- General applicable knowledge and knowledge specific to an area (e.g., aerodynamics for aircraft). Such various knowledge bases need to be integrated with one another and the application context so as to lead to a product that fulfills its functions.
- Knowledge specific to the context and necessary to assure integration between knowledge areas, as well as between these areas and the application context. This knowledge often remains tacit (e.g., transferring detailed knowledge about production process to individuals in charge of designing equipment).

Since the objective of the project front end is to define and clarify possible solutions, the integration of mobilized knowledge is to be performed among knowledge areas and the application context of the project.

8.2. Control Systems Based on Risk

Traditional management control systems were designed to master recurring organizational activities. Such control systems conventionally rely on compliance with standards and and operate in accordance with predetermined performance standards. Specific methods of project management control (e.g., the earned-value method) are essentially aimed at controlling project expenses once specifications are perfectly fleshed out. Now a project front end's objective is precisely to determine specifications. Accordingly, the main role of project front-end control systems, referring back to Galbraith's analysis (1973), is to decrease uncertainty and to increase information-processing capacity. Front-end control systems can be considered interactive in the sense of Simons (1987). Consequently, they are not aimed at complying with any predefined plan but rather supporting the project front-end phase directly.

8.2.1. Interactive Control Systems Directed to the Decrease in Uncertainty

When it comes to providing guidance, the project front end distinguishes itself from the project proper by a specific performance management method. The project front end is managed by stop-or-go decisions, although once a decision is made to undertake a project, activities are managed by adjustment. The classical distinctions in project management control relate to the difference between cost-controlled projects and profit-controlled ones:

- In the framework of a cost-controlled project, specifications, resource, deliverables, budget, organization, and payment schedule result partly from the contract negotiated with the project owner.
- In profit-controlled projects, specifications, budget, and deadlines are defined according to an environment forecast.

In both types, the role of project front end is to prepare such elements and show to the parent organization that the project can create value. This involves identifying project costs and profits, as well as incurred risks and uncertainty related in estimates, and proposing a financial model that is a synthesis of such information. Accordingly, the role of financial guidance is, in fact, mostly to supply information required to reduce uncertainty inherent in this phase of a project (Gautier, 2003). Such analysis relies on Galbraith's work (1973), as well as on Tushman and Nadler's (1978), on the role of information and of uncertainty in organizational design, and on a broad concept of management control as proposed by Simons (1990, 1995) and in the literature on Japanese target-costing practices (Tani, 1995).

On the basis of Galbraith's analyses (1973), Tushman and Nadler (1978) proposed that management control systems properly constitute efficient devices to manage uncertainty. Simons's analysis (1990, 1995) distinguishes programmed control systems, which resemble a traditional management control model leaning on a unique feedback loop, from so-called interactive control systems whose role is to gather information on strategic uncertainty (i.e., contingencies or uncertainty likely to jeopardize or invalidate a strategy) in order to stimulate the search for new opportunities and learning.

Interactive control systems fulfill three roles:

- A reporting role: When the decision-making process is vague, which is the case within design teams, interactive control systems can provide the values and preferences of management to individuals taking part in decision.making.
- A monitoring role: They guide agents by indicating the type of information to be gathered.
- A role of ratifying decisions: Interactive control systems inform managers about decisions made that engage an organization and its resources.

However, the interactive character of management control systems conceptualized by Simons (1990) is based mostly on vertical communication between subordinates and managers. In the framework of a project front end, communication within a project team, especially regarding collective settlement of problems and negotiating compromise, is also looked for. Therefore, control lies in information shared through vertical and horizontal interactions among various participants.

In analyzing the implementation of simultaneous engineering practices in Japanese industry, Tani (1995) emphasizes the importance of control systems that foster sharing information among various participants in a project. In particular, target-costing management systems form real interactive control systems. The analysis of simultaneous engineering practices in Japanese businesses provides two main lessons:

- Drastic cost cutting cannot be successful unless there is significant cooperation among various functions; advantages of simultaneous engineering are effectively obtained only when information is shared.

- Cooperation among various business divisions is important for strategic ideas to emerge. Tani (1995) notes, for instance, that design reviews constitute "in vivo" sessions of interactive control during which essential information about clients' needs or technology is shared in order to adjust the strategic design and development plan of a new product.

Consequently, interactive management control conceptually possesses a vertical dimension and also a horizontal one:

- Vertical interactive control, as conceptualized by Simons (1995), helps direct communication of objectives to a project front-end team and to get information this team possesses, thus promoting adjustment of a project's strategy.
- Horizontal interaction is at the root of integration of knowledge and information among various agents of a project front end. This interaction is designed to synthesize knowledge about a project's potential to create value and about risks pertaining to estimates. Accordingly, financial language provides a common language that integrates various perspectives and synthesizes the work done by a project front-end team (as Nixon, 1998, emphasized).

Project front-end financial guidance seems to be widely interactive, promoting the discussion between project front-end agents and managers from the parent organization. Such financial guidance aims to create a financial model expressing a project's potential value according to objectives of the parent organization.

8.2.2. The Multiple Contributions from Management Control Systems

The contributions of management control systems during the project front-end phase can be grouped into two levels.

At the decision-making level, a project front end leads to a decision whether to continue a project. Decision-making information is crucial to a parent organization because it communicates the value likely to be created by a project. This information states the risks pertaining to this estimate of this value. As we have shown in the framework of new-product design and development, a tool that is based on the principle of product cost and life cycle and that uses random Monte Carlo simulation can model risk specific to a project and its expected profitability (Gautier, 2003). Monte Carlo simulation can model risks specific to a project's design and development (Hertz, 1968). It is a technique of rational knowledge processing, especially of implicit and subjective knowledge of business experts in an organization. Business experts participate in gathering and processing information about a project from the earliest phases. During the project front-end phase, risk analysis is inseparable from assessment of the project's potential value.

Epistemologically, financial theory teaches that a decision to invest in a risky asset is based on analysis of expected return versus the risk to assets, which is measured by the standard deviation of possible profits. In the context of a unique project, risk related to events specific to the project is obviously not diversifiable, and hence managers must pay special attention to it.

According to a praxeological perspective, managers draw a distinction between a game of chance in which risks are exogenous and uncontrollable and risk taking in which information and capacities may decrease uncertainty (March & Shapira, 1987). Under this rationale, risk analysis is the basis for a risk control plan in the course of a project and thus contributes to its success.

At the organizational level, financial planning during a project front end is part of the process for preparing and designing a project. As literature on design emphasizes it (Brown & Eisenhardt, 1995), such preparatory and design work relates to four major classes of activity: problem resolution, planning, communication, and apprenticeship. Each of these classes of activity corresponds to specific research trends, which proposed particular performance factors concerning design and development activities for new products. A more thorough analysis of project front ends points out that these four classes of activity remain relevant to this particular project phase (Khurana & Rosenthal, 1998).

For problem resolution, the project front-end phase is original because its goal is to prepare a project. In this sense, as Lenfle and Midler (2003) understandably remark, a project front-end phase is for stating problems rather than solving them. This remark is all the more important given that problem settlement during the course of a project is strongly constrained by the way in which problems have been stated. A financial-based orientation implies that agents of an organization, business experts in particular, expose knowledge they possess upstream of a project. Such knowledge enables problems to be stated in technological, industrial, or economic terms.

For planning, a financial-based orientation proposes the value creation of a benchmark system by which a budget can be established as soon as a project starts. Information gathered during a project front-end phase ensures an economic-based orientation of a project when it starts.

For communication activities, an economic-based orientation relies on information that none of an organization agents possesses alone—indeed, no single agent within an organization does possess the whole information on which a financial model can be based. This information may be about clients and competitors, technical and industrial information, and resources likely to be mobilized during a project phase. A project front end based on a financial model relies on a significant social dimension that expresses itself through confidence and solidarity among protagonists (Midler, 1996). Project front-end results are not realized by the juxtaposition of subjective opinions by various organizational specialists, but by a compromise negotiated among these various specialists. Economic guidance may foster dialogue among subject experts and negotiated compromise based on financial considerations, risks, and impact options that are discovered during a project's front end.

The project front end constitutes an important phase of knowledge integration. This knowledge integration is made possible by negotiated compromise among functional experts, which results from communication and also a process of mutual learning (Hatchuel, 1994). Thus, the cooperation method in situations of collective design is one of mutual prescription, in which "each agent will let the other ones know about the prescriptions they need to comply with, in order that their speeches be compatible and result in such or such overall performance." Now, as Hatchuel (1994) emphasizes, this process of mutual prescription has to be characterized by compatibility and truth tests in order to converge. Experimentation constitutes one modality of these tests of mutual prescriptions compatibility. As far as financial considerations and risk are concerned, a financial-based model stating the value and the risks at the end of a project's front-end phase is a type of compatibility test for explorative spaces and mutual service provisions.

8.3. Conclusion

Not much interest in the project front end and guidance has been exhibited in the project management literature. However, the project front end often appears to be crucial to project success. This statement invites us to consider what may be referred to as the "prehistory of a project," during which ideas emerge in an organization, as well as objects are prepared. The project front end, which is a defining phase of a project, implies the option for a view broader than is typically acknowledged in project management. Project front-end analysis is inseparable from an organizational approach to project management because a project front end is conducted by the parent organization, which indicates close coordination with the permanent structure of the parent organization. Such an organizational view also highlights that the goals of a project front end are broader in scope than in the project management literature: Launch a project and do another project front end, or give up the very idea of the project. Performance of a project front end is not limited to project achievement but also includes specific knowledge produced during the project's front end. Many questions concerning the project front end remain unanswered: How shoud one organize a project front end? What are the specific management methods? The following question seems essential to us: How can one ensure the transition between a project front end and the project? All of these questions highlight that the avenues of enquiry into the project front end remain widely open.

References

Andersen (2008). *Rethinking project management, an organizational perspective.* Harlow, UK: Prentice Hall.

Berliner, C., & Brimson, J. A. (Eds.) (1987). *Cost management for today's advanced manufacturing, the CAM.I conceptual design.* Boston: Harvard Business School Press.

Bouquin, H. (2003). *Comptabilité de gestion.* Paris: Economica.

Bower, J. L. (1970). *Managing the resource allocation process: A study of corporate planning and investment.* Boston: Harvard Business School Press.

Brown, S. L., & Eisenhardt, K. M. (1995). Product development: Past research, present findings, and future directions. *Academy of Management Review, 20*(2), 343–378.

Cooper, R. G., & Kleinschmidt, E. J. (1987). Success factors in product innovation. *Industrial Marketing Management, 16*(3), 215–223.

Cooper, R. G., & Kleinschmidt, E. J. (1996). Winning businesses in product development: The critical success factors. *Research Technology Management, 39*(4).

Courtot, H. (1998). *La gestion des risques dans les projets.* Paris: Economica.

ECOSIP, sous la direction de V. Giard & C. Midler. (1993). *Pilotages de projet et entreprises, diversites et convergences.* Paris: Economica.

Galbraith, J. R. (1973). *Designing Complex Organizations.* Reading (PA): Addison-Wesley.

Garel, G., Giard, V., & Midler, C. (2004). *Faire de la recherché en management de projet.* Paris: Vuibert FNEGE.

Gautier, F. (2003). *Pilotage économique des projets de conception et développement de produits nouveaux.* Paris: Economica.

Gautier, F., & Giard, V. (2001). Vers une meilleure maîtrise des coûts engagés sur le cycle de vie lors de la conception de produits nouveaux. *Comptabilité Contrôle Audit, VI*(2), 43–75.

Gautier, F., & Lenfle, S. (2004). L'avant-projet: Définition et enjeux. *Faire de la recherche en management de projet.* Vuibert FNEGE, 11–34.

Giard, V. (1991). *Gestion de projets.* Paris: Economica.

Giard, V. (2003). *Gestion de la production et des flux* (3rd ed.). Paris: Economica.

Giard, V., & Pellegrin, C. (1992, March-April-May). Fondements de l'évaluation économique dans les modèles économiques de gestion. *Revue Française de Gestion,* 18–31.

Gilbert, P. (1998). *L'instrumentation de gestion—La technologie de gestion, science humaine?* Paris: Economica.

Hatchuel, A. (1994, June-July-August). Apprentissages collectifs et activités de conception. *Revue Française de Gestion,* 109–120.

Hertz, D. B. (1968). Investment policies that pay off. *Harvard Business Review, 46*(1), 96–108.

Iansati, M. (1998). *Technology integration, making critical choices in a dynamic world.* Boston: Harvard Business School Press.

Khurana, A., & Rosenthal, S. R. (1997). Integrating the fuzzy front end of new product development. *Sloan Management Review, 38*(2), 103–121.

Khurana, A., & Rosenthal, S. R. (1998). Towards holistic front ends in new product development. *Journal of Product Innovation Management, 15*(1), 57–74.

Lenfle, S., & Midler, C. (2003). Management de projet et innovation. In P. Mustar & H. Penan (Eds.), *L'encyclopédie de l'innovation.* Paris: Economica.

Lorino, P. (2001). *Méthodes et pratiques de la performance, le pilotage par les processus et les compétences* (2nd ed.). Paris: Editions d'Organization.

March, J. G. (1991). *Décisions et organizations* (translation). Paris: Éditions d'Organization.

March, J. G., & Shapira, Z. (1987). Managerial perspectives on risk and risk taking. *Management Science, 33*(11), 1404–1418.

Midler, C. (1993). *L'auto qui n'existait pas. Management de projets et transformation de l'entreprise.* Paris: InterEditions.

Midler, C. (1996). Modèles gestionnaires et régulations économiques de la conception. In G. de Terssac & G. Friedberg (Eds.), *Coopération et conception.* Toulouse: Octares Editions.

Monden, Y., & Sakurai, M. (1994). *Comptabilité et controle de gestion dans les grandes entreprises japonaises.* Paris: InterÉditions (translation of *Japanese Management Accounting,* Cambridge: Productivity Press, 1989).

Mustar, P., & Penan, H. (2003). *L'encyclopédie de l'innovation.* Paris: Economica.

Nixon, B. (1998). Research and development performance: A case study. *Management Accounting Research, 9*(3), 329–355.

Nixon, B., Innes, J., & Rabinowitz, J. (1997). Management accounting for design. *Management Accounting (London), 75*(8), 40–41.

PMI (Project Management Institute) (2008). *A guide to the Project Management Body of Knowledge® (PMBOK® Guide)—Fourth edition.* Newtown Square, PA: Author.

Simons, R. (1987). Accounting control systems and business strategy: An empirical analysis. *Accounting, Organizations and Society, 12*(4), 357–374.

Simons, R. (1990). Strategic orientation and top management attention to control systems. *Strategic Management Journal, 12*(1), 49–62.

Simons, R. (1995). *Levers of control: How managers use innovative control systems to drive strategic renewal.* Boston (MA): Harvard Business School Press.

Tanaka, M. (1994). Le contrôle des coûts dans la phase de conception d'u nouveau produit. In Y. Monden & M. Sakurai (Eds.), *Comptabilité et controle de gestion dans les grandes entreprises japonaises.* Paris: InterÉditions.

Tani, T. (1995). Interactive control in target cost management. *Management Accounting Research, 6*(4), 401–414.

Tushman, M. L., & Nadler, D. A. (1978). Information processing as an integrating concept in organizational design. *Academy of Management Review, 3*(2), 613–624.

Chapter 9

Lessons Learned from Sports Climbing: Some Disrespectful Discourse on Project Planning

Valérie Lehmann

9.1. At the Foot of a Cliff

"To read the path" is a common expression in sports climbing. Its means a climber creates and visualizes, only with the eyes, a way to access the summit of a rock face previously unconquered. This is a unique adventure project (Boudès, 2006), which is always considered a personal achievement and is generally exiting, motivating, and stressful (Gällstedt, 2003).

All climbing enthusiasts perform this type of reading at the foot of a mountain before they dash up (Boga, 1994). They know this procedure saves time, energy, anxiety, and hesitation during an ascent. Overall performance is improved, which creates efficiency.

René Caissy, a climbing instructor, has the habit of saying to his climbing students: "The aim of this exercise is to help you to plan a useful strategy for climbing, without interruption, in a fluid way. Of course, you will be surprised by the difference between your projected ascent and your realized ascent. This gap comes from your limited knowledge about your available resources and your gestural possibilities. Nevertheless, to read the path is an asset, even if it represents only a mental help."

On site, "to read the path" is a cerebral exercise that is a multifactorial analysis. To completely plan an ascent requires considering numerous variables that include the quality of the stone, weather, shadows and lighting, the length of the path, the angle of inclination, weight, size of the grips, number of cracks, placement of feet, stretching, slope, and placement of fingers. Because it is very hard to anticipate the full effects of gravity on one's own body in a move, this planning can only serve as a guideline (Jolivet, 2003).

Still, it is still not as easy as that.

Indeed, from the first step on the rock, feet and hands have to operate according to plan. Very soon "in real life" it is difficult to follow the anticipated choices, which were made for an innovative project at its beginning, when knowledge is poor and freedom is extreme (Giard & Midler, 1993). A grip may surprise when touched; perhaps it is little sticky, far less than expected. Another one gives way under weight, although it certainly looked solid. A passage, analyzed as tricky, turns out to be very easy to cross. A support promising a rest in perfect balance turns out to be total discomfort.

Better than that, a short time after leaving the ground, it is necessary to build an "other" road different from the one mentally visualized because of developments during the ascent. To think and move simultaneously quickly becomes a necessity (Schön, 1994). Here, it is urgent to stabilize in a grip, and then with eyes riveted toward the summit, to reflect a short moment before starting again. There, a dash into a crossing, which was previously unforeseen, now becomes inescapable. A fast mental scrum (Messager-Rota, 2009) is then the best means to go beyond the crux of the path.

9.2. An Irrational Move

A new approach of the cliff takes shape, emerging painfully in the mind. Also, it is suddenly necessary to move very, very fast, otherwise hands are going to fail and a fall will follow inexorably (Porter, 1986).

After a while, it becomes clear, as it has become clear to all other climbers, that the feet rather than the head should manage the ascent. This appears to be an excellent strategy: Action guides the steps up and does so with a lot of "intelligence" (Crawford, Morris, Thomas, & Winter, 2006). This special event, called the "moment of truth," is always the crux of a project, even if planning has been well done (Courtot, 1998; Gautier, 2003).

It is interesting to note that during new ascents, uncertainty results not only from the external context (e.g., temperature, wind, rock, or sponsors), as is the case with innovative projects, but also from such individual "internal" elements as the degree of motivation, anxiety, fatigue, or excitement in the face of the unknown. Also, collective "internal" factors such as the fit among team members can decrease or increase the degree of uncertainty. Thus, to implement a plan appears clearly to be just as much a matter of subjective phenomena as of objective phenomena (Midler, 1998).

Sometimes, the choice of grips is irrational. Because of the impression of a specific vertical face and the sensation that the rock is crisp, a strange and illogical step is sketched, which may prove to be worthwhile or an error.

During a climbing experience, climbers often hear a small voice that murmurs, "No, you cannot make this move. You are going to slide for sure." How many times a climber's mind plays such tricks—sometimes to the point that the climber changes the project. Moreover, "to read the path" is sometimes impregnated with this sceptical mental frame and so, it is a doubtful and negative mind that presides over tactical choices and blocks identifying many opportunities for making the ascent (Morgan, 1989).

From the very first step, during any attempt to reach a summit, climbers undergo, without any way to guard against them, the classic effects of perception. They first arise during the reading of the path. This grip here is estimated to be too small for the feet; another seems to be more welcoming for the hands; this one looks uncomfortable for fingers; and this one seems badly placed. Proud of possessing such perspicacity, the climber hurries to record mentally these analyses and take into account these crucial elements in designing the initial project (Andrews, 1971). Once suspended in the air and wandering over the rock, however, things may suddenly appear very different. The environment is not such as was evaluated: "*La paroi déjà verticale se redressa devant lui*" ("The rock face, already vertical, rose up before him"), wrote Frison-Roche (2001) in his well-known book, *Premier de cordée (The First Ascent)*. How many analyses made *a priori* are so unreliable! Every specialist assigned to a project knows that the territory is not the map (Lehmann, 2010). What is data worth compared to interpretation? Is action everything (Daft & Weick, 1984)?

9.3. The Team before Anything Else?

In a moment of peace during the ascent, tense legs and lax arms, suddenly comes the realization that this ascent's very pleasant climbing companion exercises a real power over the project without having to lead it. This partner holds the other at the end of the rope, and this link unites the two definitively during the ascent. Yes, this climbing partner retains *de facto* an indisputable authority (Bellenger, 2004) What if he decides to leave the rope without warning? Very certainly, as in most projects, stakeholders during a project do not turn out to be what they seemed at the kickoff (Jepsen & Eskerod, 2009) or during risk analysis (Bourne, 2006).

Morever, every nonhuman creature, as in the example of Callon & Law's (1989) famous scallops, is a stakeholder. The tiny, buzzing mosquito or the crazy wasp are terrible interlopers whose real power on the project only becomes clear during the ascent. Their role cannot be foreseen, even by a proven normative model. No matter which handbook or experts were consulted before departure, it is in the here and now that the real deal takes place, as it does in any project. Who said that a project was only planning, planning, and planning (Declerck, Debourse, & Declerck, 1997)?

During the ascent, the "rope companion" provides much advice that may be helpful or not. It quickly becomes clear that this companion, who can be so reassuring and helpful, is a direct partner even if he does not share the same vision for this rock climbing project. So, it is a pair who decides the way up and not just one climber. Forget the dream of a solo ascent!

This situation is even more evident when a team includes three or four climbers, especially to the climbers located in the middle of the roped party. The rope is then a link "for life and death" and thus arises a "grounded" notion of the team. Indeed, in mountaineering, as in any innovative project, the team is the place of all possibilities and all dangers (Lehmann, 2010).

To summarize this category of team life, the team always moves forward at the speed of the slowest member. So, resolution of conflicts, construction of compromises, and "creative negotiation" are common managerial activities (Midler, 1998; Asquin, Falcoz, & Picq, 2005; Picq, 2005).

Collective learning, however painful it can be, generally improves a project and enriches real-life experience (Boutinet, 1998; Vaaland, 2004). From *ad hoc* interaction emerges a stronger and smarter project (Senge, 2000), in the particular case of a climbing project of ascent from which everyone returns. Is not every project a "long chain of negotiations" (Murtoaro & Kujala, 2007)?

Finally, at the summit, a glance behind and tracing the route actually climbed reveals the real path, which seems very acceptable and satisfying. Even if the actual project looks little like the one the climber envisaged during project planning, it corresponds fundamentally to the climber's personality, skills, and resources gathered along the path (Avenier, 1997). It is time to ask if the same pattern is not used in all projects. That is, elegantly jumping from grips to grips under the gaze of admirers, or taking grips as quickly as possible, always urgent to finish the challenge at hand, and applauded by colleagues. Who has tried to understand the success of projects carried out exclusively using organic resources (Jaafari, 2003; Andersen, 2006)?

9.4. Crossing the Golden Triangle

Definitively, any ascent plunges a climber into situations that require a show of determination. This determination is needed to keep feet and hands from letting go. It is needed to command the energy needed to overcome any paralyzing fear of the void below. Certainly, leadership helps every team member to feel more comfortable with the twists and turns of a project (Müller & Turner, 2007), but each team member must also learn to distrust emotions that can play terrible tricks (Sotiaux, 2008). For example, a subjective danger appears along on the path. This risk is minor, but it is perceived as serious because of both its possibility of occurrence and its grave impact on the ascent; on the contrary, a really high risk may be perceived as insignificant (Courtot, 1998). No matter the risks, Pinto would doubtlessly say that R&D projects are only "spontaneous" realizations (Pinto, 2002).

All climbers have experienced the sensation of fear beyond reason that urges them to modify their initial strategy (Cox & Fulsaas, 2003). For example, a climber rejects a small grip, which is in reality very safe, for a larger grip that demands much more physical effort from the climber. Even the best climbers do not control totally their fear of falling. Reaching a summit is often made at the price of each climber's heart skipping too many beats. Emotion, not always "intelligent," knows how to strike just as a climber reaches the big space, in the silence of the summit (Laforêt, 2004). And at the end, project objectives can seem so faint.

When the ascent is finished, exhausted and overtaken by an intense experience of planning and climbing, the initial purpose has hardly any importance in itself. Nothing else remains for a climber, except for a comforting sensation of satisfaction. Never after a challenge that has been met and accomplished has there been such a feeling of peace. The climber is ready to swear: "But why did I read this path before leaving? I would have been happier to live that experience without any formal planning and to trust completely my body and my soul."

When complexity occurs, most specialists in project management have experienced this desire of immateriality (Jolivet & Navarre, 1993). Some people call this state "playing God." In mountaineering, this strange feeling is known as "small death." It is like a drug, which, unfortunately, has prevented numerous climbers from returning home. It may be another way of striking one's wall of incompetence. It is similar to what project managers experience when they rely solely on the "golden triangle" (i.e., cost, quality, and time) to succeed (Kerzner, 2003).

By the way, climbing with eyes shut can be an immense pleasure even for novices. When climbing this way, there is neither purpose nor means. There are only feet and hands that caress the cliff and are like a blind person's cane. Could all projects take advantage of such an "agile" approach?

References

Andersen, E. S. (2006). Toward a project management theory of renewal projects. *Project Management Journal, 37*(4), 15–30.

Andrews, K. R. (1971). *The concept of corporate strategy* (3rd ed.). New York: Richard D. Irwin.

Argyris, C., & Schon, D. (1978). *Organizationl learning: A theory of action perspective.* Reading, MA: Addison Wesley.

Asquin, A., Falcoz, C., & Picq, T. (2005). *Ce que manager un projet veut dire.* Paris: Éditions d'organisation.

Avenier, M-J. (1997). Stratégie tâtonnante et démarche projet: Une modalité née dans le contexte des opérations de construction publiques. In M. J. Avenier (Ed.), *La stratégie chemin faisant* (pp. 269–298). Paris: Economica.

Bellenger, L. (2004). *Piloter une équipe projet.* Paris: ESF Éditeur.

Boga, S. (1994). *Climbers: Scaling the heights with the sport's elite.* Mechanicsburg, PA: Stackpole Books.

Boudès, T. (2006). Le déroulement d'un projet envisagé comme un récit d'aventure. In O. Germain (Ed.), *De nouvelles figures du projet en management* (pp. 37–52). Paris: Editions EMS, Management & Société.

Bourne, L., & Walker, D. H. (2006). Using a visualising tool to study stakeholder influence–Two Australian examples. *Journal of Project Management, 37*(1), 5-21.

Boutinet, J.-P. (1998). Le management par projet et logique communicationnelle: Quelles convergences, quels défis? *Revue Communication et Organisation* (Montréal: GRECO), *13,* 52–60.

Callon, M., & Law, J. (1989). La proto-histoire d'un laboratoire ou le difficile mariage de la science et de l'économie. In M. Callon et al., *Innovation et ressources locales.* Paris: PUF.

Courtot, H. (1998). *La gestion des risques dans les projets.* Paris: Éditions Économica.

Cox, M., & Fulsaas, K. (2003). *Mountaineering.* The freedom of the hills. Seattle, WA: The Mountaineers, Don Graydon.

Crawford, L., Morris, P., Thomas, J., & Winter, M. (2006). Practitioner development: from trained technicians to reflective practitioners. *International Journal of Project Management, 24,* 722–733.

Daft, R. L., & Weick, K. E. (1984). Toward a model of organizations as interpretation systems. *Academy of Management Review, 9*(2), 284-295.

Declerck, R. P., Debourse, J. P., & Declerck, J. C. (1997). *Le management stratégique, Contrôle de l'irréversibilité.* Lille, France: Éditions ESC.

Frison-Roche, R. (2001). *Premier de cordée.* Paris: Collection J'ai lu, Édition Poche.

Gällstedt, M. (2003). Working conditions in projects: Perceptions of stress and motivation among projects team members and project manager. *International Journal of Project Management, 21,* 449–455.

Garel, G., Giard, V., & Midler, C. (2004). *Faire de la recherche en management de projet, ouvrage collectif coordonné par.* Paris: Éditions Vuibert.

Gautier, M. (2003). *Pilotage économique des projets de conception et développement de produits nouveaux.* Paris: Éditions Économica.

Giard, V., & Midler, C. (1993). *Pilotages de projet et entreprises, diversités et convergences.* Sous la direction de Collectif ECOSIP. Paris: Éditions Économica.
Jaafari, A. (2003). Project management in the age of complexity and change. *Project Management Journal, 34*(4), 47–57.
Jepsen, A. L., and Eskerod, P. (2009). Stakeholder analysis in projects: Challenges in using curent guidelines in the real world. *International Journal of Project Management, 27,* 335–343.
Jolivet, F. (2003). *Manager l'entreprise par projets.* France: EMS.
Jolivet, F., & Navarre, C. (1993). Grands projets, auto-organisation, méta-règles: Vers de nouvelles formes de management de grands projets. *Revue Gestion 2000, 93*(2).
Karlsen, J. T. (2002). Project stakeholder management. *Engineering Management Journal, 14*(4), 19–24.
Kerzner, H. (2003). *Project management: A systems approach for planning, scheduling, and controlling.* Hoboken, NJ: John Wiley & Sons.
Laforêt, Y. (2004). *L'Everest m'a conquis,* nouvelle édition. Montréal: Édition Stanké.
Lehmann, V. (2010). Connecting changes to projects using a historical perspective: Towards some new canvases for researchers. *International Journal of Project Management, 28*(4), 328–338.
Messager-Rota, V. (2009). *Gestion de projet, vers des méthodes agiles.* Paris: Éditions Eyrolles.nce
Midler, C. (1998). *L'auto qui n'existait pas.* Paris: Éditions d'Organisation.
Morgan, G. (1989). *Images de l'organisation.* Montréal: Presses de l'Université Laval.
Müller, R., & Turner, R. (2007). Matching the project manager's leadership style to project type. *International Journal of Project Management, 21,* 8–13.
Murtoaro, J., & Kujala, J. (2007). Project negotiation analysis. *International Journal of Project Management, 25,* 722–733.
Picq, T. (2005). *Manager une équipe de projet* (2nd ed.). Paris: Éditions Dunod.
Pinto, J. K. (2002). Project management 2002. *Research Technology Management, 45*(2), 22.
Porter, M. E. (1986). *L'avantage concurrentiel.* Paris: InterÉditions.
Schön, D. A. (1994). *Le praticien réflexif.* Formation des Maîtres (version originale 1983). Paris: Éditions logiques.
Senge, P. (2000). *Schools that learn: A fifth discipline fieldbook for educators, parents, and everyone who cares about education.* New York: Crown Business.
Sotiaux, Y. (2008). *Management d'équipe projet: mode d'emploi.* Paris: Gereso.
Vaaland, T. I. (2004). Improving project collaboration: Start with the conflicts. *International Journal of Project Management, 22,* 447–454.

Chapter 10

Managing Extreme Situations in Fire and Rescue Organizations: The Complexity in Implementing Feedback

Anaïs Gautier

Feedback (REX) has long been the subject of research in the fields of risk management and knowledge management. It is defined as learning from the study of an event or phenomenon in order to fully understand the mechanisms leading to malfunction or innovation. The feedback process corresponds to a producer of individual and organizational learning from an event and is naturally a tool of risk management. According to many authors, feedback is a risk-related study of experience to ensure the reliability of a tool, a system, or an organization. It has the characteristic of not studying a part of a phenomenon according to intentions, hierarchical positions, and skills of actors. It often boils down to technical analysis, which is usually not a general look at systems behavior but more of a focus on the organizational environment. Feedback approaches in reliability are not suitable for a fertile and favorable expansion in industry organizations and in public institutions. The technical approach has not allowed a development of process feedback to a change in a learning organization (Senge, 2001).

This shift has occurred by authors developing other design feedback processes within organizations and institutions (Gilbert & Bourdeaux, 1999; Lagadec, 2001; Gilbert, 2001; Wybo, 2002, 2006; Lecoze, Lim, & Dechy, 2002; Van Wassenhove, 2004; Hadj Mabrouk & Hadj Mabrouk, 2004; Gaillard, 2005; Van Wassenhove & Garbolino, 2008; Duret & Lassagne, 2008; Gautier, Lièvre, & Rix, 2008; Lièvre & Gautier, 2009). There are two main concerns. The first is about formalization for feedback by the diversity and evolution of different approaches (e.g., technical, human, organizational, and contextual), and the second focuses on the potential development of organizational learning enabled by the application. Theorists have developed methods and tools that are scientific and highly impractical for the uninitiated. We believe feedback is a tool that actors can formalize and appropriate under a language that is specific to their jobs and situations that they face and alone know in detail from experience. However, they must be "accompanied" to ensure organizational learning. It is necessary to highlight the need for a multidisciplinary approach and clinical feedback in organizations. Toward this objective, we propose and attempt to implement and develop feedback in a dimension for organizational learning. Our feedback design is dynamic and is focused on managing natural, daily situations (Bourrier, 2001; Journé, 2005) rather than the accidental character of specific events. We favor a systemic approach to situations in which organizations operate because situated learning is a process that occurs in stages: identifying gaps between actions and situation (i.e., perception error and individual learning), learning collectively in a single or double loop depending on the nature of the gaps identified (Argyris & Schön, 2002), observed findings and lesson products, and then applying this collective learning to an organizational learning strategy for sustainable change. We examine a variety of issues in this chapter. We look at the origin of this work in an article written by Weick (1993). This article also motivated us to choose the empirical investigation of a rescue and fire service. Then we continue with the conditions necessary for implementing feedback in organizations facing extreme situations of management to expose the characteristics and particularities of this research topic.

10.1. Origins of Our Research Project and Focus on Analysis of Management Situations in Fire and Rescue Services

The origin of our work on organizational learning, and organizational context is related to an article by K. E. Weick (1993) on the tragedy of Mann Gulch (1949), in which 13 members of a team of 16 smokejumpers were killed. This article made many contributions to the theory of organizations, including the notion of sensemaking in a situation. We are also interested in qualitative, in-depth analysis close to the action of subjects in a small group (e.g., an intervention group) in extreme situations (e.g., forest fire). Weick's article influenced our choice for fire and rescue service and, more particularly, forest fire situations.

10.1.1. Sensemaking Applied to Management Situations

The sensemaking of the Mann Gulch tragedy involves methods for retrospectively analyzing the management of situations. If sensemaking emerges primarily in a specific context of management situations (Journé & Raulet-Crozet, 2004, 2008) and particularly extreme situation management (Lièvre, 2005; Gautier et al., 2008; Lièvre & Gautier, 2009), it is an indispensable tool in the functioning of organizations in operation. In our study, we integrate the concept of strategic experiential logistics (Lièvre, 2007), which considers experienced situations as knowledge mobilized into action. This is similar to Weick (1995), who considers that action takes precedence over cognition. The concept of strategic experiential logistics allows actors to be learners capitalizing on knowledge produced by the experience of everyday situations. This is a topic that we discuss in connection to our choice of organization for this study.

10.1.2. An Organization Stigmatized by Its Own History

In this chapter, we are interested in fire and rescue services and the skills to be learned from action situations. Fire and rescue organizations in France are old institutions dating from 1790 (Dalmaz, 1998); they first appeared after the fire of Rennes in 1720, which stirred awareness of the cost to society of fire's destruction. We do not review all the exciting and tumultuous history of firefighting, but examine only two major catastrophes in the history of the French fire and rescue services.

The first is related to the Decree of 8 September 1811, which created the battalion of firefighters in Paris (BSPP). The Austrian Embassy fire had taken place the night of July 2, 1810, at the time of the marriage of Emperor Napoleon I and Empress Marie-Louise. There was great material loss that included the theft of jewelry and precious objects. Panic among wedding guests caused many injuries, and a member of the ambassador's family died. This led to a radical reform to correct the the numerous failures of the rescue and fire service. The absence of the city's guard commander and the presence of less rigorous leaders in the control hierarchy led to the reorganization of the institution and the establishment of military authority (i.e., military rules and regulations from the infantry) under the orders of the Commissioner of Police and the Minister of the Interior.

On October 28, 1938, a fire occurred in Marseille on the first floor of the popular department store, *Nouvelles Galeries.* This disaster's toll included 13 dead, 56 missing, and 22 injured rescuers. This event put France into a state of national mourning. The government created the battalion of marine brigade in Marseille and placed it under military authority. Lack of coordination and improvisation caused the rescue operation to be very complex and inefficient. The investigation by the Ministry of the Interior reported that emergency management was not effective due to the lack of firefighters stationed on alert; the lack of centralized equipment delayed rapid mobilization, and there was insufficient equipment and number of firefighters (many staff from the nearby cities were called in). This event highlighted lack of staff and equipment to meet the needs of a large city like Marseille.

This story shows that an urban population requires fire and rescue services that are efficient, reliable, and well organized and resourced. In 1938, the French people were shocked by this tragedy. The law governed the operation of the local emergency organizations but also started to regulate the safety and security in the design and operation of commercial and public buildings.

The consequences of these events and their impact on the French fire and rescue services led us to understand why this long-established institution knows no common culture and formal feedback nowadays. This is surprising for a risk organization. Numerous factors do not allow for the existence of systematic analysis review of situations. Therefore, we examine the conditions necessary for the implementation of practical feedback in the following section.

10.2. The Organizational Framework of Feedback

This chapter discusses the lessons learned in a specific context of organizations that perform fire and rescue and the operational management to reduce the risk of forest fires. This research focuses on a specific methodology to analyze events that are dynamic, unpredictable, and risky, among other characteristics. The feedback in this research study has two features. One is the nature of events and their scope, which is related to professional identities and specific practices (i.e., skills). The second is the nature of organizations whose practices are influenced by organizational culture.

Context is essential in organizing action during extreme situations. An extreme situation (Lièvre, 2005) can reveal the actions of actors and their limits. Operational circumstances are indicative of the skills used by actors to quickly understand a situation (i.e., context and task to be performed) to produce sensemaking within a very short time.

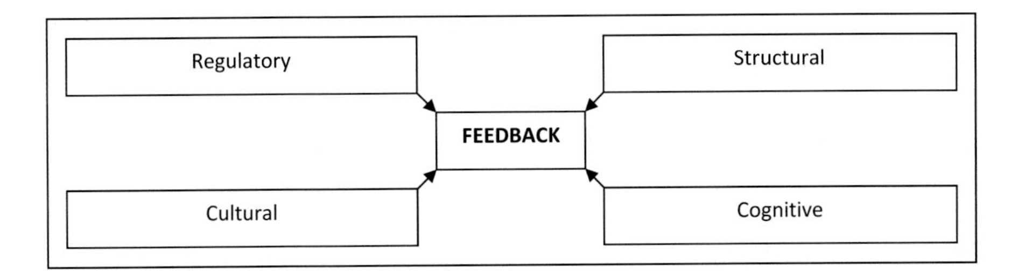

Figure 10.1 The organizational framework of feedback.

We conducted diagnostic feedback tests in fire and rescue services to understand the representations of this tool and the meaning which it gave. This work identified four factors to explain the lack of a common culture and formalized feedback (see Figure 10.1). We called this group of factors the "organizational framework of feedback" because these factors represent constraints in implementating feedback within operational practices. We discuss the various factors and how we have incorporated them into the design of our research.

10.2.1. The Regulatory Factor

The regulatory factor is the formalization of feedback in organizational processes. Formalizing feedback is necessary to extract and produce organizational learning and corrective actions. This relates to feedback as a tool for knowledge management (Ermine, 1996). It includes a principle of ownership of the concept by actors so it can fit into the rhythm of their activity. Practice of feedback should be seen as part of daily action rather than an activity to mark time or break "time out," which is unbearable for actors accustomed to emergencies. Feedback must be part of the operational activity and not appear different from tactical action. To implement normative feedback, we insisted that it be recognized as a function of expertise. Recognizing feedback as a function of expertise allows it to be considered as more than using a tool. Feedback is thus integrated more easily into an operational activity and does not burden actors with an additional task. Regarding feedback as an operational function makes it fully a matter of expertise using specific knowledge.

10.2.2. The Structural Factor

The structural factor refers to regional and local French public authorities. In French rescue and fire services, feedback is based on local initiatives and voluntary actions, which are local and autonomous. These practices owe their existence to the motivation of some actors who want to learn. These practices appear to be genuine tools to improve performance within systems, and we pose the question: What is the status of feedback in these organizations? Is feedback a method to advance the profession through organizational learning? Or is feedback simply a tool claiming more efficient fire and rescue services to prove they are among the best in practicing a form of internal self-skills? To promote common culture and formalized feedback, we developed a module of feedback initiation practices in national training specializing in forest fire fighting. The aim is to integrate procedures and knowledge about feedback. Training can establish general principles for the development of a common culture of feedback.

10.2.3. The Cultural Factor

The cultural factor concerns the beliefs, myths, values, and norms that characterize an organization. Through our observations, we find that the organizational culture did not value a process of feedback because its values are opposed to analysis of situations. These values include: the heroic, mythic status surrounding the image of firefighters and rescuers; legitimacy gained through hierarchical and social recognition of skills and experience; the emotions generated from some interventions; and the principle of a mean obligation in operations. These values are all components that remain tacit in these organizations. These principles are part of the organizational culture and are difficult to reconcile with a feedback practice. To promote a feedback practice, we chose to work on management situations (Journé, 1999; Journé & Raulet-Croset, 2008, Lièvre & Gautier, 2009, Gautier et al., 2008) and, more particularly, on situations conductive to learning. These are common situations (Bourrier, 2001) that are not subject to legal oversight and represent nearly 90% of all operational situations.

10.2.4. The Cognitive Factor

The cognitive factor is about the analysis of an actor's behavior in a situation. We consider how error is viewed in these organizations and the perception of the gap between situations and an actor's behavior in action (Desmond, 2006). This factor is similar to the concept of cognitive dissonance, which describes the differing interpretations of the same situation by various actors. Being a firefighter is a vocation and identity chosen by the majority of actors we studied; for them, it is not just a job. The human factor is related to man, who has sense and reason. The difficulty in clearly identifying behaviors and triggers is a reason for the existence of informal feedback. Feedback can be difficult when it is obvious and touches collective areas considered personal. The view of error becomes important in this debate. In an emergency services organization, the norm is to practice "avoidance strategy" or consider the grounds of incompetence (Desmond, 2006), and not to analyze and not to create problems or update responsibilities. The rule is not to know rather than to know what one does not want to see [i.e., denial of reality in the sense of Roux-Dufort (2004)]. To circumvent this, when we saw a noncompliant behavior in a situation, we tried to understand the origin of this "gap" by using a feedback model consisting of four registers. These registers are sometimes obstacles but are a prerequisite for the learning process (Gautier et al., 2008). This model allows the highlighting of the behavior of actors with the operation of the organization and makes them aware of the "gap" without stating explicitly that an error was committed. This promotes learning from real situations and avoids misunderstandings and any negative conduct.

10.3. Conclusion

This analysis of implementing practice feedback has highlighted four factors that limit a culture of feedback in fire and rescue services.

1. The regulatory factor implies integration into daily activity, because feedback must be a formalized practice integrated into operational activity to promote the process of sensemaking.
2. The structural factor concerns the organizational structures that limit communication and information sharing.
3. The cultural factor involves the identity aspects of an organization. Military values of heroism appear difficult to reconcile with a feedback practice, which sometimes questions values, beliefs, and models.
4. The cognitive factor involves cognitive processes that make collaboration and feedback difficult because feedback can raise awareness of actions, which may be at odds with the organizational values.

In fact, these factors appear to limit the implementation of practical feedback focused on organizational modes. Awareness of these factors is a prerequisite to any implementation of a formalized practice feedback, and it is the first step in a broader study of the place and role of feedback in organizations at risk.

To conclude, the formalization and learning potential of feedback are central. Feedback is within an organizational framework that is defined by four factors, which represent constraints of organizational modes. Therefore, any discussion prior to implementing feedback must take into account these factors and integrate them into overall practices for feedback and learning within organizations.

References

Argyris, C., & Schön, D. (2002). *L'apprentissage organisationnel: Théorie, méthodes, pratiques.* Éditions De Boeck Université, France.

Bourrier, M. (dir.) (2001). *Organiser la fiabilité.* Paris: L'harmattan.

Dalmaz, P. (1998). *Histoire des sapeurs-pompiers français* (2nd ed.). Presses Universitaires de France, Paris.

Desmond, M. (2006). Des morts incompétents. *Actes de la recherche en sciences sociales, 4*(165), 8–27.

Duret, R., & Lassagne, M. (2008). *Communautés de pratiques et gestion du retour d'expérience: Le cas d'une entreprise de construction.* Colloque AIMS 2008, Nice, France.

Ermine, J.-L. (1996). *Les systèmes de connaissances.* Paris: Editions Hermès.

Gaillard, I. (2005). Etat des connaissances sur le retour d'expérience industriel et ses facteurs socioculturels de réussite ou d'échec. *Cahier de L'ICSI,* no. 2005-02.

Gautier, A., Lièvre, P., & Rix, G. (2008). Les obstacles à l'apprentissage organisationnel au sein de la sécurité civile: Une mise en perspective en terme de gestion des ressources humaines. *Revue politique et management public, 26*(2), Paris.

Gilbert, C. (2001). Retours d'expérience: Le poids des contraintes. *Annales des mines responsabilité et environnement: Recherches débats et actions,* no. 22.

Gilbert, C., & Bourdeaux, I. (dir.) (1999). Retours d'expérience, apprentissages et vigilance organisationnels. Approches croisées. *Actes la cinquième séance,* 24 mars 1999, Paris.

Hadj Mabrouk, A., & Hadj Mabrouk, H. (2004). Approche d'intégration de l'erreur humaine dans le retour d'expérience. *Cahier de la recherche, INRETS,* 1 février 2004.

Journé, B. (1999). *Les organisations complexes à risque: Gérer la sûreté par les ressources. Etudes de situation de conduite de centrales nucléaires.* Thèses de doctorat, Ecole Polytechnique, Paris, 434 pages.

Journé, B. (2005). Etudier le management de l'imprévu: Méthodes dynamiques d'observations in situ. *Finances Contrôle Stratégie, 8*(4), 63–91, Paris.

Journé, B., & Raulet-Croset, N. (2004). Le concept de situation dans les sciences du management: Analyse de l'incertitude, l'ambiguïté et l'imprévu dans les organisations, *Association Internationale de Management Stratégique (AIMS),* 13° Conférence AIMS, Normandie-Vallée de Seine France, 3–4 juin.

Journé, B., & Raulet-Croset, N. (2008). Le concept de situation: Contribution à l'analyse de l'activité managériale dans un contexte d'ambiguïté et d'incertitude. *Management, 11*(1), 27–55.

Lagadec, P. (2001). *Retour d'expérience: Théorie et pratique, Le rapport de la commission d'enquête britannique sur l'encéphalopathie spongiforme bovine (ESB) au Royaume-*

Uni entre 1986 et 1996. Cahiers du GIS risques collectifs et situations de crise, no. 1, Publication de la MSH-Alpes.

Lecoze, J.-C., Lim, S., & Dechy, N. (2002). *Intégration des aspects organisationnels dans le retour d'expérience, l'accident majeur un phénomène complexe à étudier*. Etude DRA-16, INERIS.

Lievre, P. (2005). *Vers une logistique des situations extrêmes! De la logistique de projet du point de vue d'une épistémologie de l'activité d'une expédition polaire* (pp. 11–20). HDR, Université Aix-Marseille II, France.

Lievre, P. (2007). *La logistique*. Éditions La découverte, Collections repères (gestion), France.

Lievre, P., & Gautier, A. (2009). Les registres de la logistique des situations extrêmes: Des expéditions polaires aux services d'incendies et de secours. *Revue Management & Avenir*, numéro spécial Piloter des supply chains: Quels enjeux inter-organisationnels et réticulaires?, no. 24, 2009/4, 196–216.

Roux-Dufort, C. (2004). La performance, anti-chambre de la crise. *Cahiers de recherche (Working papers)*, no.2004/04, EM Lyon.

Senge, P. (2001). *La cinquième discipline: Le guide de terrain*. Paris: Editions Générales First.

Van Wassenhove, W. (2004, December). *Définition et opérationnalisation d'une organisation apprenante (OA) à l'aide du retour d'expérience, application à la gestion des alertes sanitaires liées à l'alimentation*. thèse de doctorat de l'ENGREF (bio-industriel).

Van Wassenhove, W., & Garbolino, E. (2008). *Retour d'expérience et prévention des risques, Principes et méthodes*. Collection SRD Sciences du risque et du danger, série Notes de synthèse et de recherche, éditions Tec et Doc (Lavoisier), France.

Weick, K. E. (1993). The collapse of sensemaking in organization: The Mann Gulch disaster. *Administrative Science Quarterly*, *38*(4), 628–652.

Weick, K. E. (1995). *Sensemaking in organizations*. Foundations for organizational science, Sage publications.

Wybo, J.-L. (2002). *Apprentissage organisationnel à partir de l'analyse de la gestion de risques technologiques et naturels*. Les rencontres AMRAE 2002, Table ronde Risques industriels et alimentaires: De la théorie à la pratique.

Wybo J.-L. (2006). *Mémento sur la conduite du retour d'expérience*. Document interne du Ministère de l'Intérieur et de l'Aménagement du territoire, Direction de la Défense et de la Sécurité Civiles, Sous-direction de la gestion des risques, Bureau de l'analyse et de la préparation aux crises.

Chapter 11

Coordination Practices in Extreme Situations: Lessons from the Military

Cécile Godé

Over the last decade, several contributions have suggested that refining our understanding of intra-organizational coordination requires opening the "black box" of coordination processes (e.g., Berman, Down & Hill, 2002; Faraj & Xiao, 2006; Becky, 2006; Klein, Ziegert, Knight, & Xiao, 2006; Rico, Sanchez-Manzanares, Gil, & Gibson, 2008; Okhuysen & Bechky, 2009; Lechner & Kreutzer, 2010). Indeed, in reckoning coordination as essentially a matter of structure, the dominant perspective of contingency (e.g., March & Simon, 1958; Lawrence & Lorsch, 1967; Galbraith, 1973; Mintzberg, 1978; Gupta, Dirsmith, & Fogarty, 1994) fails at exploring the micro level of coordination, as well as in delivering a comprehensive appreciation of the way individuals handle collaborative tasks in a situation. Such theories draw on the assumption that an environment remains sufficiently predictable to allow prespecification of the most efficient modes of coordination. Moreover, these theories favor studying coordination at the organizational level, and they do not examine how coordination is performed by individuals in practice.

Because of these limitations, understanding how coordination occurs under extreme situations remains interesting to investigate. Drawing on Girin's work

(1990), extreme situations refers to management situations that are at the same time evolving, uncertain, and risky (Bouty et al., 2012; Godé, 2016). The evolving nature of extreme situations emphasizes rapid, discontinuous, and simultaneous change (Faraj & Xiao, 2006; Wirtz, Mathieu, & Schilke, 2007). The uncertainty of extreme situations refers to the probability of events occurring, as well as the moment of occurrence and practical details. Finally, extreme situations entail a high level of risk that can be vital, symbolic, or both.

This chapter questions the opportunities for learning lessons by examining the way teams coordinate through their ongoing actions (Johnson, Melin, & Whittington, 2003) when they operate under extreme situations. More specifically, it explores coordination practices developed by fighter aircrews and ground forces during war operations. Investigation of military teams' activities reveals that combat fighters know exactly the objective to perform from the start, but often have to decide how the mission should be run during the course of action. Due to unstable conditions, they frequently challenge the usual coordination processes, formulate others, and implement new solutions within the existing framework of local rules and routines.

The chapter proceeds as follows. Section 11.1 develops a military case study, which concerns coordination practices of fighters during air-to-ground operations in Afghanistan. Because of the high instability with which these experts cope, they continuously have to create and re-create coordination to accomplish the strike missions in which they are involved (Godé, 2010). Section 11.2 discusses theoretical implications of the case analysis. Section 11.3 is the conclusion.

11.1. Coordinating in Extreme Situation: The Case of Air-to-Ground Operations in Afghanistan

11.1.1. Case Setting and Methodology

In Afghanistan, French forces were involved in two distinct operations that were running in parallel:

- Enduring Freedom was a U.S.-led coalition action supporting counterterrorism after the events of 9/11/2001.
- The International Security Assistance Force (ISAF) was intended to bring stability in Afghanistan and prevent the emergence of terrorist cells in the region.

Through these operations, the main coalitions goal was to pass Afghanistan's security over to the Afghan people.

Within this context of war, the French fighter aircraft Mirage 2000D was engaged in air-to-ground missions called "close air support" (CAS). CAS consisted of air action against hostile targets that were in close proximity to friendly forces. War fighters acted under time-sensitive pressure and hostility because they were targeted by enemy ground fire. As an element of joint fire support, each service organized CAS within its role as part of the joint force. As a result, CAS required perfect coordination between ground and air forces. Usually, a forward air controller (FAC) led the action of fighter aircraft from the ground by transmitting the appropriate information over the radio. The FAC communicated the precise target location and ensured that aircrews understood the situation on the ground. The FAC was the most qualified service member to perform such activities and assumed all the responsibilities associated with targeting. In that way, coordination challenges occurred at two levels: between the pilot and the navigator, which was the aircrew; and between the aircrew and the FAC, who guided the airplane when it arrived above the combat area. To investigate how combat fighters resolved coordination issues in high-speed environments, this extreme single-case study (Yin, 2003) was used to explore the way fighter aircrews and forward air controllers coordinated in practice.

A mix of data collection methods was used to achieve triangulation (Eisenhardt, 1989). Data sources included:

1. Individual and collective interviews
2. Observations from shadowing
3. Archival records and reports from the field

Further, it is critical to consider that the researcher's department was located on a French Air Force base. As a result, on a daily basis, the researcher had opportunities to collect information through informal conversations, as well as to observe to deeply understand the military community, its codes, languages, and practices.

Individual and Collective Interviews

Eleven individual, semistructured interviews were conducted with four pilots, five navigators, and two forward air controllers. All of them had been recently in Afghanistan. Narration was encouraged to understand how they coordinated in action, under stressful, hostile, and time-sensitive conditions. The goal was to identify coordination practices through narratives of combat experiences. Further, a collective interview was conducted with members of the Mirage 2000D squadron based in Kandahar. The interview sought to stimulate interaction among squadron members to highlight problems of task division, teamwork, and coordination. The same guide was used to conduct individual and collective interviews. It was developed to reflect three main themes:

- Environmental characteristics
- Coordination processes and elements of coordination used in practice
- Coordination failures

The objective was to collect enough data to be able to characterize a war context and understand what routine and unpredictable events are for a combatant, and to describe with details the mechanisms, means, and tools exploited by combatants to coordinate.

Observations through Shadowing

Finally, observations from shadowing two professional meetings allowed the author to grasp the reality of coordination practices referred to during the interviews. Observations were made using a flight simulator regarding the way aircrews work (e.g., communication practices, artifacts used, nonverbal behavior) to launch a bomb.

Archival Records and Reports from the Field

French archival records and reports from the field were gathered. They were about the aircrews and included FACs' action reports on mission improvement and potential coordination problems that were dealt with. Also studied were newer video and audio records (in particular, dialogues between aircrews and FACs during strike actions).

Coordination Practices Developed by Warfighters to Perform Close Air Support in Afghanistan

Standardized Language and Automatisms as Basic Coordination Elements in Extreme Situations

At the aircrew level, tasks were precisely divided between the pilot and the navigator according to their competences and expertise. Usually, the navigator had to manage medium-term and long-term tasks (e.g., electronic threat monitoring, radio frequency control, and weapons tracking), while the pilot was focused on flying and firing, which are short-term activities. A navigator explained:

> For a pilot, short term is the next two, three minutes: intercepting a fighter aircraft, for example, or reacting to an engine failure and immediately starting up the engine again. Whereas medium and long term concern all the activities taking more than two or three minutes to be achieved. That is the navigator who works with the FAC to find the target on the ground while the pilot manages his patrol, for example.

Execution of CAS operations is covered by rules of engagement (ROEs). These are directives issued by competent military authorities that delineate circumstances and limitations under which forces have to conduct combat engagement. Moreover, during the entire mission, aircrews and FACs interact according to strict procedures that include air tasking orders (ATOs), checklists, and CAS cards. In that way, they work within a highly standardized context. A pilot explained: "Standardized? It means that we use well-known models to respond to each stage of the mission."

How was a routine CAS mission conducted? According to the time schedule in the Air Tasking Order, the pilot reached the FAC by radio in order to confirm primary and secondary radio frequencies, identification friend or foe codes, and NATO authentications. Once the aircraft had been authenticated, the FAC passed target information on to the aircrew from a standardized document called the CAS card. The CAS card is made up of data lines indicating the enemy formation and disposition, the time in minutes and seconds for the aircraft to fly from the initial point to the target, the exit route for the aircraft after the attack, and other information. Right after the CAS card stage, the aircraft flew to the target area. At this time, another FAC began to describe the environment to the navigator to gradually allow him to visually identify the target without any ambiguity.

Together with standardized rules and procedures, combat fighters used a common language, published within NATO documentation—what they called "code words." A pilot stated:

> Code words are meaningful. They represent a strong base for a common language. You don't have to interpret, you don't have to think. For example, "investigate" means that you must identify your target. We know immediately that we must make a particular kind of interception to observe and identify. No ambiguity. Code words represent a kind of communication philosophy.

The meanings of code words were well known among the combat fighters. A single word enabled them to share a large amount of information accurately and very quickly.

This point highlights the critical role played by automatism to coordinate in extreme situations. A navigator explained: "Automatisms are habits which emerge from aircrew work. In fact, aircrew life is quite similar to living as a couple! You make odd habits!" Indeed, once in theater, aircrews always flew together (they are called "constituted aircrews" or "war aircrews"). This provided an opportunity to develop internal synergies. A pilot pointed out: "Working together enables us to internalize standard action patterns, which are maneuvers and dispositions. These allow us to build our representation of the aerial space. How could I say? We train together; we have intuitive combat between us. It is like a football team. We have experiences and training in common." Automatisms facilitated collective sensemaking building and enabled aircrews to produce a common interpretation of a tactical situation. Moreover, they played a critical role in reducing internal communication, because each aircrew member knew how to interpret the others' actiond and what to do in response. Automatisms played an essential role in providing combat fighters with a stable structure that was not questioned in the course of action. Automatisms enabled combat fighters to save time, which they could allocate to other activities. A pilot pointed out: "When a pilot copes with an unexpected situation, he saves precious time thanks to automatisms that he can use to reckon more alternative courses of action, for example."

Mutual Trust and Collective Creativity as In Situ *Coordination Elements in Extreme Situations*

Being reactive and adaptive is decisive in handling extreme situations. FACs and aircrews must be able to quickly rebuild some sense of what is happening to avoid suddenly losing meaning, which could result in disaster. More precisely, pilots and navigators frequently used the time saved by automatisms to communicate; they had to agree on appropriate solutions that could be placed into practice to achieve their goals. When doing this, they tended to abandon the use of code words and employ a natural language. A pilot told us:

> In theater, stress can be very high. In some circumstances, your objective is to stay alive. During these times, communication is the most important, no matter the language you use.

Another pilot added:

> Even if we've learned a very standardized language, we tend to adopt a more intuitive and everyday one during combat. . . . In fact, internal dialogue between pilots and navigators increases as soon as the

situation becomes awkward. Our goal is to pass information on, no matter the means.

Instead of communicating using standardized coded language, they had discussions using everyday language. In doing so, they exchanged their opinions concerning the specifics of the unfolding event and used their creativity to find an appropriate solution. For instance, if friendly and enemy forces were in close proximity and ground combat was very intense, the aircrew could decide not to engage the target in order to avoid fratricide. Instead, they could imagine different solutions, as in a show of force such as "buzzing" the area, which is a very-low-altitude flight intended to intimidate the Taliban. A navigator related an event he was involved in: "Last year, we decided to not fire. . . . there were Rules of Engagement, what we were seeing on the ground. . . . All together, we debated the opportunity to engage the target, and we finally agreed not to." In these circumstances, natural language is selected because interactions within very stressful environments tend to drive people to fall back to action and communication patterns they had learned previously and more fully (Weick, 1990). Debates and discussions are efficient ways to produce creative solutions on what to do to coordinate when an unexpected event occurs.

Under such intense pressure and hostility, mutual trust among combat fighters is critical, because survival depends on their ability to perform tasks and remain focused. Doubts about others' competencies are not allowed. Combat fighters have to take on a great responsibility in their course of action, and trust is able to ensure that they share a common knowledge regarding their goals and capabilities. A pilot pointed out: "Trust is a kind of mutual protection insurance." Another went deeper in explaining that collective performance depends on trust: "With Captain X, we take off and we are good. How can we do that? Just because we've known each other for a long time now. We share a mutual knowledge and we trust each other." Interpersonal trust helps to turn stress into positive emotions and to stimulate team performance.

Such necessary trust emerges from social relationships, both professional and friendship. For flight crews, sustaining these relationships is cultural. The aircraft culture, shaped by the reality that the Mirage 2000D is a two-seat aircraft, and a squadron's habits represent critical sources of interpersonal trust. A navigator indicated: "A squadron is like a tribe. It's become cultural to trust each other." As noted previously, in theater, aircrews always fly together; they are constituted. Another pilot explained: "When we are in theater, patrols and aircrews are made, and we never change team members. It is like a tribe." Such arrangements leverage interpersonal trust because each flight crew gets to know others and to assess the way they work, as well as their ability to adapt to changing circumstances. FACs and aircrews also seek to develop interpersonal trust because they know that trust is the basis for collective work in extreme situations. A FAC underlined: "FACs' proximity with squadrons is the key to a good job. It reduces their lack of aeronautic culture. Moreover, it allows aircrews to trust in FACs." Another went deeper: "Having interpersonal relationships—it's a part of our job. It's necessary to be curious and interact with pilots, know their procedures and be in line with what they do." French Air Force pilots, navigators, and FACs share a common philosophy of what their job must be.

11.2. Coordinating in Extreme Situation through Bundles of Practices

Our case study shed new light on coordination theory underlying the combinative nature of coordination. We observed that combat fighters bundle practices of coordination to deal with uncertain, changing, and risky events by continuously using articulate mechanisms, means, and tools of coordination (Table 11.1).

To handle coordination in extreme situations, combat fighters first develop bundles of coordination practices based on standardization of work practices and language. Moreover, they are able to apply suites of procedures and actions, which they call "automatisms," that can be viewed as general patterns of interaction interiorized by team members and that structure their behavior. This is particularly the case for aircrews; they share a stock of common knowledge accumulated over time though mutual training and experience. That allows them to automatically anticipate the actions and needs of their colleagues and to adjust their own behavior. This situation describes an implicit coordination process (Rico et al., 2008) and highlights actors' ability to produce a collective result on the basis of a set of highly standardized practices and devices.

Table 11.1. Bundles of Coordination Practices

Extreme Situations: Uncertain Changing and Risky Conditions	
Coordination mechanisms "The glue holding organizational structures together" (Mintzberg, 1978, p. 19)	Standardization of work processes, standardization of outputs, mutual adjustment
Coordination means Techniques used by individuals to reach their goals	Rules, procedures, standardized language, automatisms, natural language, culture, trust, consensus, social relationships
Coordination tools Devices supporting coordination	Air Tasking Order, checklists, code words, Procedures, radio, dialogue, informal face-to-face

However, standardization and automatisms are not sufficient for coordination in extreme situations. The case study also underlines the critical role played by trust and creativity in coordination.

By having mutual trust, team members perceive that interactions with each other are safe and easy. In other words, they draw mainly on emotional and social relationships developed over time to build a collective sense and produce new combinations of coordination *in situ*. They have to be creative when pressure and stress are intense, and they do not have the time to question their colleagues' competencies. Being creative in such circumstances requires sharing the same interpretation of the environment and applying a collective sense to the events they face. Collective sensemaking allows the management of task interdependencies to be carried out in the context of relationships. In that way, it can be regarded as a prerequisite because it helps actors to creatively recombine elements of coordination.

Based on the case of French combat fighters deployed in Afghanistan, and their experience in close air support missions, we inferred two main managerial outcomes that should be taken into consideration by civilian organizations:

- Socialization processes
- Turnover policy

First, combat fighters' feedback recurrently outlined the importance of what they called *cohesive activities,* and which in fact related to a set of managerial means enabling the increase of convergence and the accumulation of team members' tacit knowledge over time. In effect, when people share extra-professional time together, they express their desire to be or to remain part of the team and, what is more, they get to know each other. In that way, a team with a high level of cohesion is usually considered more effective in converting time spent interacting together into valuable collective knowledge. In addition, cohesion allows teams to be more motivated to participate actively in team processes (Berman et al., 2002). The French Air Force is particularly attentive to such team needs and seriously takes combat fighters' feedback into consideration. As a result, different solutions have been implemented in theater. For instance, a well-known context of socialization was reproduced and adapted in Afghanistan: the squadron bar. This was a place where pilots and navigators, and occasionally other combat fighters involved in aerial missions, got together and talked freely about their day, the pressure they had felt, the situation they had been involved in, the solutions they had implemented, the errors they had made, and other topics. In other words, the squadron bar was a place where team members gradually built a stock of common knowledge, providing them with a shared and accurate understanding of what their work and their roles were within the team.

Such a common knowledge is critical for creativity, because, in this case, combat fighters could convert it into valuable resources once they had coped with unexpected events.

This illustrates that though individual competencies are required to perform teamwork, they are not sufficient to provide efficient results. Team performance depends significantly on the ability to develop a collective sense of events and interdependency and to use common knowledge and attitudes. These elements constitute collective team competencies, representing more than simply the sum of the individual competencies and expertise of team members (Salas, Goodwin, & Burke, 2009). We observed that the team/task environment determines the processes by which collective competencies are achieved. To this extent, in enabling team members to be creative and innovative, socialization practices play a critical role.

Another managerial outcome learned from combat fighters' practices and relevant to civilian organizations concerns turnover policy. It is difficult to turn a collective stock of knowledge into practice when teams are affected by high turnover. The ability of team members to draw on experientially constructed sensemaking depends on the time they spend together. Team efficiency is crucially dependent on its stability. However, it is also important to take into consideration the risk of "knowledge ossification" (Berman et al., 2002, p. 14), outlining that significant levels of team experience are able to produce negative returns after some point in time. It is possible to observe a decline in team performance, in which core competencies turn into core rigidities. Again, the French Air Force appears to be attentive to such issues. In theater, combat fighters are deployed for just a few months (between two and three months). As a result, there is not enough time to experiment with the effect of knowledge ossification. During this short period in theater, combat fighters can take advantage of different kinds of socialization processes (such as a squadron bar) to quickly increase the common stock of knowledge. We observe that at both the organizational and team levels, centripetal and centrifugal forces operate in combination to manage the tensions in learning between coordination and turnover (Godé & Bouty, 2011).

11.3. Conclusion

In investigating the way military teams coordinate during air-to-ground missions in Afghanistan, our study has produced results that can be consequential for civilian organizations' activities. Military teams have to coordinate with each other in handling a kind of "tension" between the autonomy requirement, which cultivates creativity, and organizational standards and routines.

Observing the whole of coordination practices and managerial solutions implemented by these military teams, we inferred two main outcomes that could be easily adapted and disseminated in organizations. These concern socialization processes and turnover policy, which play a critical role in enabling an appropriate team-task environment.

References

Becky, B. (2006). Gaffers, gofers, and grips: Role-based coordination in temporary organizations. *Organization Science, 17*(1), 3–21.

Berman, S. L., Down, J., & Hill, C. W. (2002). Tacit knowledge as a source of competitive advantage in the National Basketball Association. *Academy of Management Journal, 45*(1), 13–31.

Bouty, I., Godé, C., Drucker-Godard, C., Nizet, J., Pichault, F., & Lièvre, P. (2012). Coordination practices in extreme situations. *European Management Journal, 30,* 475–489.

Eisenhardt, K. (1989). Building theories from case study research. *Academy of Management Review, 14*(4), 532–550.

Faraj, S., & Xiao, Y. (2006). Coordination in fast-response organizations. *Management Science, 52*(8), 1155–1169.

Galbraith, J. R. (1973). *Designing complex organizations.* Reading, MA: Addison-Wesley.

Girin, J. (1990). L'analyse empirique des situations de gestion: Eléments de théorie et de méthode. In A. C. Martinet (ed.), *Epistémologie et sciences de gestion* (pp. 141–182). Paris: Economica.

Godé, C. (2010). Leveraging coordination in project-based activities: What can we learn from military teamwork? *Project Management Journal,* Special Issue: Project management in extreme situation, *41*(3), 69–78.

Godé, C. (2016). *Team coordination in extreme environment: Work practices and technological uses under uncertainty.* ISTE & John Wiley.

Godé, C., & Bouty, I. (2011). *Benefiting from shared cognition in coordination processes: Individual and organizational responses in an air display squadron.* Conference on Coordination within and among Organizations, ASQ OMT Division of the AOM–HEC, June 13–14.

Gupta, P., Dirsmith, M., & Fogarty, T. (1994). Coordination and control in a government agency: Contingency and institutional theory perspectives on GAO audits. *Administrative Science Quarterly, 39*(2), 264–284.

Johnson, G., Melin, L., & Whittington, R. (2003). Micro strategy and strategizing: Towards an activity-based view? *Journal of Management Studies, 40*(1), 3–22.

Klein, C., Ziegert, J., Knight, A., & Xiao, Y. (2006). Dynamic delegation: Shared, hierarchical and deindividualized leadership in extreme action teams. *Administrative Science Quarterly, 51,* 590–621.

Lawrence, P., & Lorsch, J. (1967). *Organization and environment.* Boston: Harvard Business School Press.

Lechner, C., & Kreutzer, M. (2010). Coordinating growth initiatives in multi-units firms. *Long Range Planning, 43,* 6–32.
March, J., & Simon, H. (1958). *Organizations.* New York: John Wiley.
Mintzberg, H. (1978). *The structuring of organizations: A synthesis of the research.* Englewood Cliffs, NJ: Prentice-Hall.
Okhuysen, G., & Bechky, B. (2009). Coordination in organizations: An integrative perspective. *Academy of Management Annals, 3*(1), 463–502.
Rico, R., Sanchez-Manzanares, M., Gil, F., & Gibson, C. (2008). Team implicit coordination processes: A team knowledge-based approach. *Journal of Management Review, 33*(1), 163–184.
Salas, E., Goodwin, G., & Burke, C. (2009). *Team effectiveness in complex organizations: Cross-disciplinary perspectives and approaches.* New York: Psychology Press.
Weick, K. (1990). The vulnerable system: An analysis of the Tenerife air disaster. *Journal of Management, 16*(3), 571–593.
Wirtz, B., Mathieu, A., & Schilke, O. (2007). Strategy in high-velocity environment. *Long Range Planning, 40,* 293–313.
Yin, R. K. (2003). *Case study research: Design and methods.* Thousand Oaks, CA: Sage.

Chapter 12

Developing Collective Competence in Extreme Project Teams: The French Special Forces Case

Tessa Melkonian and Thierry Picq

Developing work groups able to carry out projects in extreme situations that are evolving, uncertain, and risky (Lièvre, 2005) is a considerable challenge for a growing number of organizations. Until very recently, managing extreme situation was important only to very specific organizations, such as nuclear power plants (see Gauthereau & Hollnagel, 2005). However, the question of managing extreme situations is now crucial for more classic organizations working in a socioeconomic environment that also combines complexity, uncertainty, and risk (Berry, 2005). However, we still know very little about the concrete collective mechanisms at work in projects carried out in extreme situations (Faraj & Xiao, 2006), especially in regard to collective competence.

This chapter looks at the experience of the French Special Forces whose specificities are the setting up of groups focused on action in a hostile, politically sensitive, and uncertain context. First, we define and discuss the concept of collective competence and its relevance in the specific context of project management. Then we briefly present the Special Forces and the unusual methodology we used to study it, and we provide elements to understand the components of collective competence. Finally, we discuss the importance of considering both pre- and postproject processes to better understand the development of collective competence in extreme action teams.

12.1. Conceptual Framework: Developing Collective Competence in Teams

12.1.1. The Concept of Collective Competence

Interest in collective competence in companies has grown stronger and stronger because of the development of organizations based on group work and the increasing number of cross-disciplinary collective entities, which have a variety of forms (e.g., matrix structures, task forces, project teams, networking, and shared practices). A similar interest in the concept can be observed in the academic world (Doz, 1994; Krohmer & Retour, 2004; Zarifian, 1999). Recent attempts to approach the notion of collective competence conceptually are part of the vast literature in human resources management (HRM) that focuses on "the competence model" (Zarifian, 1995). Indeed, a number of authors attribute to collective competence some characteristics already attributed to individual skills. In fact, those collective skills are observed and developed in action, where they are defined in relation to a problem situation. However, research also recognizes that collective competence is more than just the simple addition of individual skills. According to Dejoux (1998), collective competence is the collection of individual skills plus an indefinable component that is unique to the group. Collective competence therefore remains a "black box" to be explored.

Practitioners and academics have thus to contend with the issue of answering conceptual challenges specific to a composite notion situated at an intermediate level, which is between the individual and the organization levels. They also have to contend with the operational challenge of giving a new boost to a more collective and cooperative dimension of work in the face of the limitations noticed in the increase of HRM individualization.

12.1.2. Collective Competence in Project Mode

Projects are now widely used in companies of every size and sector, and they seem to be particularly suited to the application of the competence model. If one defines a project as "a plan adapted to its subject" (Garel, 2003), the question of assembling the most appropriate skills is key for the success or failure of the project. However, as Midler (1993) pointed out, the fundamental characteristic of a project is its collective dimension. The ability to manage effectively a project team composed of actors with distinct but complementary skills, specialties, experience, and profiles, and sometimes even different national and cultural backgrounds, requires knowing how to go beyond the mere addition of individual skills. The strength of a project stems from an adequate combination of skills that achieves an objective, which is unachievable with isolated skills. A project therefore appears to be a particularly appropriate context in which to study the nature and dynamic of this "strange attractor" that is collective competence (Le Boterf, 1994). A summary of the research that focuses specifically on the construction of collective competence in projects highlights two principal levels of analysis:

1. That of the processes through which collective skills are developed in projects
2. That of the concrete measures that facilitate the emergence of these processes (for a review, see Dameron, 2002; Garel, 2003; Loufrani-Fedida, 2006)

12.1.3. Studying the Development of Collective Competence: Issues and Questions

Our intention is therefore to observe, analyze, and model the development process of collective skills in the context of extreme project teams. Although the concept of collective competence is promising, especially for modern organizations, it nonetheless raises a number of operational questions that concern the identification of components, evaluation, development, management, and links with performance. On a more conceptual basis, the notion of collective competence raises the key question of isomorphism (Fillol, 2004), in other words, interactions between different levels of analysis (e.g., individual, collective, and organizational) and cognitive, affective, organizational, and relational mechanisms at work in the dynamics of interpersonal synergy.

In the tradition of the work of Bataille (2001), we chose to tackle the recursive individual and collective relationship using a constructivist approach that stresses the emergence of collective competence in a complex time dynamic (Guilhon & Trépo, 2000). Our research question is the following: How are collective skills formed and developed over time in projects in extreme situations? To answer this question, we chose to study the development of collective skills in a very specific project context that offers both a connection with the time conducive to the observation of dynamic and progressive phenomena and that possesses particularly marked "extreme" characteristics: that of the commandos of the French Army's Special Forces (SF).

12.2. Project Management in Extreme Situations: The Case of the Special Forces

12.2.1. A Brief History

The SF came into being in Great Britain during World War II in the battle against Nazi Germany. Commando units were a unique response to the overwhelming dominance of the enemy. Their daring strikes occasionally reversed the sense of victory by the enemy. Resorting to unconventional actions to surprise the enemy and instill doubt is still very much at the core of the values that sustain the SF. The know-how developed during World War II was maintained and then refined during the wars in Indo-China and Algeria. However, the end of these colonial conflicts, replaced by the Cold War between the United States and the Soviet Union, pushed these special units back into the shadows. During the Gulf War, the need arose again for light, very mobile, and well-armed units made up of highly trained team members. The British resorted to this type of commando action to neutralize the Scud missiles directed against Israel. In France, the inhibitions linked with the Algerian conflict were gradually overcome. The associations with the excesses of counter-guerrilla warfare were replaced by the rediscover of Special Forces core capabilities, which had been undervalued by decision makers. France created a Special Operations Command (COS) in 1992, after the first Gulf War. It exercises its operational authority on the SF units of the three armies (Army, Marine, and Air Force). Today, only France, Great Britain, and the United States have such units.

12.2.2. Specificities of the SF Commando Projects

The capacity to train teams that are both cohesive and effective is the dominant organizational trait of the SF. These are elite units that put very mobile means into operation to carry out targeted missions that have high added value. These missions can last a few hours or weeks in a context that is particularly hostile, complex, and uncertain. When exposed, these operations have considerable political implications, so they require extreme discretion and a quality of execution that is equal to the strategic stakes in play. The specificity of these units is to bring together potentially very diverse areas of expertise that can include explosives, transmissions, information, optics, topography, and sharpshooting. In interservice missions, these skills are augmented with a combination of skills that link ground, sea, and air components. Not only does each individual member have to be a specialist in his field, he also must be capable of combining his expertise with that of the others to benefit the mission. The need for an optimal combination of skills is reinforced by the limited number of team members, which often numbers between five and ten, and by the fact that, once engaged in the action, a commando team maneuvers completely autonomously in hostile surroundings.

A commando intervention is organized around two main stages. The first one is devoted to seeking strategic information, and the second stage uses this information to carry out the action itself (e.g., arresting war criminals, freeing hostages, or neutralizing a terrorist group). In the context of the French FS, these two elements (i.e., information and action) are divided between two distinct regiments: the Thirteenth Dragoon Parachute Regiment and the First Marine Infantry Parachute Regiment. This reinforces the need for each unit to develop adjustment and cooperation skills.

12.2.3. Methodology and Specific Conditions of Study

Our main research objective was to understand how collective skills were developed in commando projects in the past. We therefore wanted to question SF commandos still in service and those not in service. Given the *official secret* aspect of the activity, we had to send a formal request for authorization to the Special Operations Command (SOC), as well as to the Command of the Land Army, explaining that the subject was not going to be about the content of the missions. Once we had received official authorization, we met several commandos in service on their base. Recording the conversations was strictly prohibited. To try and limit the loss of data because of minimal note taking, all the interviews were systematically carried out by two researchers.

In accordance with recommended processes to develop theory, we carried out our qualitative approach inductively and in two stages (Glaser & Strauss, 1967). Our first stage was broad and aimed at understanding how the SF commandos operate, by familiarizing ourselves with their history, their vocabulary, and their operational specificities. We thus carried out open interviews with commandos in service and former members of the SF. We completed this first stage of data collection through secondary sources, in particular through specialized work on the Special Forces (notably Dénecé, 2002). The second phase included semistructured interviews with SF members of varying levels of hierarchy and responsibility (e.g., young commandos, team leader, unit leader, regiment commander, and human resources management). In the next section of this chapter we present the main themes that emerged during this research.

12.3. Development and Mobilization of Collective Competence in the Special Forces

12.3.1. *Strong Individual Expertise*

The analysis of the SF teams is a reminder that the construction of collective competence relies above all on strong individual expertise that is recognized and suitable for a particular mission. This individual quality is built up as a continuous process, throughout a long career path full of selection points and commando training. The initial training begins with a four-week pretraining course, which is very trying both physically and mentally, and at the end of which only a small percentage of candidates is retained. There follows an intensive training course for about ten weeks. Once he has integrated into his unit, a commando has to acquire specific qualifications (e.g., operational free-fall parachutist, marksman, radio, or marker). During a career period in the Special Forces, these specialized skills are reinforced by continuous training in sessions that last from a half-day to several days, for a total annual amount of about one hundred days, in addition to any operations. Psychological aspects are as important as the technical aspects in this training. These psychological aspects include resistance to stress (i.e., interrogation simulation and daring), survival capacity (i.e., ability to escape and survive in hostile surroundings), resistance to surroundings and bad weather, and team spirit instilled through persuasive teaching. As soon as a mission order is issued by a political authority, the SOC establishes a list of interservice requirements. It is the responsibility of each unit to supply as quickly as possible—sometimes within an hour—individuals with top expertise and who are psychologically prepared to face the unexpected. They form an *ad hoc* mission team coordinated by an officer, from information or action, depending on the type of mission.

12.3.2. *The Combination of Different but Complementary Expertise*

Having talented experts is not enough to guarantee the collective performance of a commando team. Collective performance is prepared, outside any missions, in training courses that mix interservice, interregiment and interexpertise, during which the commandos continuously exchange their experiences. The objective is to create a cross-disciplinary situation, which can combine execution methods of different forms of expertise, and which can be called on in a mission. This continuous testing through training sessions makes it possible for each individual to know how to play his role fully in the group and to learn how to develop social cohesion, confidence, and cooperative behavior quickly.

Commandos familiarize themselves with common procedures for preparing and executing mission types and to operate in deterioration mode. Thus, prepared for the worst, they increase their ability to resist stress and adapt to danger, concentrating on the precise role that they have to play and on collective reference points of every nature (e.g., elementary acts, collective sequences, and procedures) that they have learned individually during training sessions.

12.3.3. The Construction of a Shared Representation Based on Common Reference Systems and Language

Acquiring basic procedures and collective automatic reflexes represents a common ground where combinations of action specific to each mission can be developed. It is during the preparation phase of the mission that the capacity for collective situated action is constructed. There is a common interservice methodology to tackle missions based on the scenario method. This makes it possible to envisage all the possible options in a given situation. These decision trees refer to general diagrams, which are developed at the SOC level, adapted during the preparation of the mission, and then refined in the theater of operations itself. This preparation of scenarios is an essential element of the common reference system that structures the commandos' collective action. This collective action is a product of a general guidelines procedure, initial pieces of information on the precise situation, and the past experience of the actors, which leads to a common representation of the problem situation illustrated through drawings, graphs, cards, and other aids, which facilitate a common visualization. The process occurs very frequently and is participative; all team members are involved in refining procedures. They alternate between workshops on different fields of competence and mixed presentations. The team makes presentations to the mission command, which seeks out and challenges weak points in the presentations.

12.3.4. The Capacity of Collective Improvisation

Whenever conditions permit, this work is carried out in a unique physical place, which has adjustable partitions that make it possible to alternate work between small specialist groups and collective meetings. The French FS favor co-presence and direct contact, contrary to their U.S. counterparts, who resort to information and communication systems that enable distance work. The back office is at the center and provides information live on the context of the mission to help the teams refine their preparation. The coordination and control of information and action is carried out at the level of the operation captain. The captain entirely assumes the role of support interlocutor, who takes on, as much as possible, coordinating questions during a mission. In actual high-risk situations, the capacity for collective initiative and improvisation is essential. Responsibilities can thus evolve during a mission as all the types of expertise link together (i.e., action must not hinder the information). The chain of command in a team can always be readjusted according to circumstances. However, using the principle of respecting the real hierarchical chain, which is a point of reference known to all, avoids confusion and facilitates the decision-making process in extreme situations.

12.3.5. A Collective Memory

In the SF, capitalizing on experiences after each mission is a vital step, which is part of the method of organizing by project. The construction of a collective memory both traces past training and guides future training. This benefitting from experience is carried out at different moments:

1. Continuously, even during the operations, to be able to readjust to possible future actions
2. At the end of the mission, on the spot, in the form of individual or collective debriefing to get the most important information quickly
3. After a couple of weeks via a detailed team return on experience called RETEX (RETurn on EXperience).

This formal procedure results in a document and official cards based on precise data related to the operation, which may include list of times, actions, and visual aids. The purpose is to assess the execution methods for training and improvement rather than sanctioning them. Capitalizing on experiences affects the memory of operations, which can be performed again in another context, as well as the organization, the action methods, the procedures, the equipment and the preparation, and on training as a whole. In addition to supporting technical and methodological aspects, RETEX also psychologically supports team members, who may have been afflicted by events experienced during a mission. Managing the "postmission" is an essential element in protecting the human potential in this high-risk context.

12.3.6. Subjective and Shared Commitment

A final ingredient of collective competence highlighted in the SF is the subjective commitment of the actors, who use all their energy to carry out each project as a high-risk event with the extreme consequences of life or death. Developing this capacity of commitment is an integral part of training, which is made evident by the extreme rigor of individual and collective preparation for a mission. This is illustrated by the ritual of preparing equipment, which involves equipment and the personal bag. It is also illustrated during action by the complete mobilization of every mental, physical, and intellectual capacity of every team member, who is aware that an individual error could lead to collective failure and put the life of the group at stake. Finally, it is illustrated by the seriousness accorded to debriefing. The solidarity and "social commitment" created in an *ad hoc* group, whose members often did not know each other until just a few hours before the mission, relies not only on the awareness shared by all of the dangerous nature of the mission but also on the fundamental values of the French SF. The commandos know that they can count on each other, for the rule is that they must never leave one of their own behind.

12.4. Discussion: From Project Management to Management by Project

12.4.1. Contributions to Collective Competence in Projects

The study of how the commandos of the Special Forces operate contributes to a better understanding of the elements that underpin the development of collective competence in project situations. In regard to theory, our work promotes reflection on a concept that is recent and less than stable. By approaching the project as a complex area where the contributions of each person are combined, we strengthen the view that collective competence is a consequence of a complex combination of individual and specialized knowledge (Amherdt, Dupuich-Rabasse, Emery, & Giauque, 2000). We validate the importance of four components already identified in the literature on collective competence, namely, the existence of a common reference point and shared language, of a collective memory and a subjective commitment (Khromer & Retour, 2006), by describing their application in an extreme context, which reinforces its relevance. Our work also encourages a rapprochement between the literature on collective competence and the literature on projects, which has for a long time tackled the positive impact of creating a shared representation in a multidisciplinary team (Ramonjavelo, Préfontaine, Skander, & Ricard, 2006). It also shows ties to the literature on high-reliability organizations (HROs), which defends the major role of collective improvisation capability in the development of collective competence in risk situations (see, notably, Weick & Roberts, 1993).

12.4.2. Contributions to the Dynamic of the Development of Collective Competence

Our study also sheds light on a dynamic and multilevel approach to collective skills development where the pre- and postmission phases play an essential role.

The Preparation of a Mission

Preparation is a resource for action and is a cognitive support that is shared to enhance team members' improvising together. It is during preparation that each individual's specialty skills are discussed, organized, and integrated. The formulation of action scenarios and the discussion of alternative tactics, fallback plans, and other similar items generate a cognitive shared area, which is a type of common system of reference that creates an *ad hoc* interdependence essential to deploying positioned collective skills (Bataille, 2001). This development stage of a collective cognitive structure can be analyzed as the creation mechanisms of sense or *sensemaking* (see Weick, 1993). By using the foundational works of Weick (1993, 1995), we understand that the information, uncertainties, and events to be produced during a mission can only make sense and be handled in interpretative contexts, constructed and shared beforehand through discussions, exchanges, and meetings during the preparation stage. Obviously, this process does not focus only on the technical aspects. It also focuses on social aspects. As a system of interpretation governing the relations with the world and others, it directs and organizes management and communications. As collective phenomena, they involve individuals' social affiliation with the affective and normative implications (Moscovici, 1972).

Postmission Management

The SF model illustrates quantitatively and qualitatively intense learning approaches. Emphasis is on the collective (i.e., learning together) and active (i.e., learning through action) dimensions. In addition to the highly important simulated ground training sessions, which offer numerous opportunities to create knowledge in action (Midler, Boudès, & Charue-Duboc, 1997), the RETEX plays a major role in this constant search of permanent learning at individual, team, and organization levels. The importance of the return on experience is often described in the literature on HROs (Kervern, 2005) and is discussed more and more frequently in management as an important mechanism that may enable traditional organizations to maximize their organizational learning (Vashdi, Bamberger, Erez, & Weiss-Meilik, 2007). Finally, the logic of the profession, where former commandos pass on their experience during training sessions, is a lesson for all the companies that ask themselves the question of how they can use the expertise of their "seniors."

Recent work by Asquin, Garel, & Picq (2005) takes a critical view of projects and focus on the risks individuals may face in projects, which include increased pressure (in terms of work load and weight of responsibilities), individualization of career paths, loss of collective solidarity, and weakening of expertise. Here also, the SF offer proof of these risks being taken into consideration. For example, the end of a mission can be a very delicate phase. The need to communicate within a unit is very strong in this phase, because it is important for the unit to have a better understanding of each person's responsibilities and to put into perspective aspects of the mission that could create psychological damage. In this sense, the men must above all avoid keeping things to themselves just as they come off a mission. In a similar way, wounded officers and noncommissioned officers are frequently kept in a unit for as long as possible, to train newer commandos or perform operational support functions. These noncombat positions enable the SF to realize a fuller human potential, and they give those who have served a status and place worthy of their personal commitment and sacrifice. These aspects of the SF remind us that the capacity to manage the end a project is of particular importance to organizational success.

12.5. Conclusion

The premission therefore enables a shared ability for understanding, the mission uses collective skills, and the postmission, because of its introspective nature, makes learning possible. We feel that this statement is an important lesson for an organization in which action often takes it into preparation and capitalization phases. An outstanding operational lesson that the SF model offers companies is found in the SF's ability to reconcile a one-off exceptional performance for each mission while maintaining lasting efficiency that goes well beyond each mission. In today's modern company, success in a project just once is not enough. Competitive advantage is achieved through the ability to perform successfully continually, or, in other words, to generate a regular stream of effective projects.

References

Amherdt, C. H., Dupuich-Rabasse, F., Emery, Y., & Giauque, D. (2000). *Compétences collectives dans les organisations*. Québec: Presse de l'université de Laval.

Asquin, A., Garel, G., & Picq, T. (2005). Le côté sombre des projets. Gérer et comprendre. *Annales des mines, 90,* 43–54.

Barney, J. B. (2001). Is the resource-based "view" a useful perspective for strategic management research? *Academy of Management Review, 26*(1), 41–56.

Bataille, F. (2001). Compétence collective et performance. *Revue de gestion des resources humaines,* Avril-Mai-Juin, 66–81.

Berry, M. (2005). *Management de l'extrême* (Vol. 1). Paris: Editions Autrement.

Dameron, S. (2002). Les deux conceptions du développement de relations coopératives dans l'organisation. In I. Dolaster & H. Laroche (Eds.), *Perspectives en management strategique* (Vol. VIII). EMS Éditions.

Dejoux, C. (1998). Pour une approche transversale de la gestion des compétences. *Gestion 2000,* Nov.-Dec., 15–31.

Dénecé, E. (2007). *Histoire secrète des Forces Spéciales de 1945 à nos jours.* Editions Nouveau Monde.

Doz, Y. (1994). Le dilemme de la gestion du renouvellement des compétences clés. *Revue française de gestion,* no. 97(Mai-Juin), 79–91.

Faraj, S., & Xiao, Y. (2006). Coordination in fast-response organizations. *Management Science, 52*(8), 1155–1169.

Fillol, C. (2004). Apprentissage et systémique: Une perspective intégrée. *Revue française de gestion,* no. 149(Mars-Avril), 33–50.

Garel, G. (2003). *Le management de projet.* Editions La Découverte.

Gauthereau, V., & Hollnagel, E. (2005). Planning, Control, and Adaptation: A Case Study. *European Management Journal, 23(*1), 118–131.

Glaser, B. G., & Strauss, A. L. (1967). *The discovery of grounded theory: Strategies for qualitative research.* New York: Aldine de Gruyter.

Guilhon, A., & Trépo, G. (2000). La compétence collective: Le chaînon manquant entre la stratégie et la gestion des ressources humaines. In IXème Conférence Internationale de Management Stratégique, Montpellier.

Kervern, G.-Y. (2005). Histoire des cindyniques, émergence de nouveaux 'patterns': Déconstruction de la destruction. In E. Morin & J.-L. Le Moigne (Eds.), *Intelligence de la complexité: Epistémologie et pragmatique.* Cerisy.

Krohmer, C., & Retour, D. (2006, December). *La compétence collective, maillon clé de la gestion des compétences.* Paris: GRACCO, Communication à la journée de recherche, Groupe gestion des compétences.

Le Boterf, G. (1994). *De la compétence: Essai sur un attracteur étrange.* Ed. Organisation.

Lièvre, P. (2005). *Vers une logistique des situations extrêmes, de la logistique de projet du point de vue d'une épistémologie de l'activité d'une expédition polaire.* HDR, Université Aix Marseille II.

Loufrani-Fedida, S. (2006). *Management des compétences et organisation par projets: Une mise en valeur de leur articulation.* Thèse de doctorat, Université de Nice.

Midler, C. (1993). *L'auto qui n'existait pas: Management des projets et transformation de l'entreprise.* Paris: Interéditions.

Midler, C., Boudes, T., & Charue-Duboc, F. (1997). Formation et apprentissage collectif dans les entreprises: Une expérience dans le domaine du management de projet. *Revue Internationale de Gestion, 22*(3), 86–92.

Moscovici, S. (1972). *Introduction à la psychologie sociale.* Paris: Larousse.

Ramonjavelo, V., Préfontaine, L., Skander, D., & Ricard, L. (2006). Une assise au développement des PPP: La confiance institutionnelle, interorganisationnelle et interpersonnelle. *Canadian Public Administration, 49*(3), 350–374.

Senge, P. (1990). *The fifth discipline: The art and practice of the learning organization.* New York: Doubleday.

Vashdi, D. R., Bamberger, P. A., Erez, M., & Weiss-Meilik, A. (2007). Briefing-debriefing: Using a reflexive organizational learning model from the military to enhance the performance of surgical teams. *Human Resource Management, 46*(1), 115–142.

Weick, K. E. (1993). The collapse of sensemaking in organizations: The Mann Gulch disaster. *Administrative Science Quarterly, 38*(4), 628–652.

Weick, K. E. (1995). *Sensemaking in organizations* (Vol. 3). Thousand Oaks, CA: Sage.

Weick, K. E., & Roberts, K. H. (1993). Collective minds in organizations: Heedful interrelating on flight decks. *Administrative Science Quarterly, 38*, 357–381.

Zarifian, P. (1995, November). Coopération, compétence et système de gestion dans l'industrie: La recherche de cohérence. *Actes du 5ème congrès de l'AGRH* (pp. 15-20), Montpellier.

Zarifian, P. (1999). *Objectif compétence: Pour une nouvelle logique.* Paris: Liaisons.

Chapter 13

Situated Teams: Dropping Tools on Mount Everest

Markus Hällgren

Team formation is a common practice in contemporary temporary organizations (i.e., projects) whenever an organization faces significant threats or problems. Although this need is appreciated, there is less understanding of how it happens in practice. The aim of this chapter is to contribute to that understanding. Team formation is investigated through the lens of Weick's "dropping the tools," which essentially implies that an organization or an individual has to set aside everyday practices (e.g., norms, values, rules, and routines) in order to survive (Weick, 1993). The empirical basis is first-hand bibliographic accounts of events that occurred in 1996 on Mount Everest. The analysis shows that people teamed up in three different types of situated team types—task, survival, and rescue—and that they did so because of five different types of trigger: rules, goal achievement, obstacle, necessity, and expectations. The chapter offers implications for academics, practitioners, and lecturers on organizational theory in general and temporary organizations in particular.

Managing the unexpected is the greatest challenge any manager faces (Weick & Sutcliffe, 2001, p. 1). Sometimes the unexpected contributes to disasters, such as the events on Mount Everest in 1996 that killed nine (Kayes, 2004). In such temporary organizations as projects the unexpected is the rule rather than the exception (Lundin & Söderholm, 1995). Temporary organizations thus present great challenges to any individual who chooses to live and work in the context of them.

Team formation is a common practice in contemporary project organizations whenever an organization faces significant threats or problems. (Engwall & Svensson, 2001, 2004; Pavlak, 2004; Hällgren & Wilson, 2008). Although this need to form a team is appreciated, there is less understanding of how it happens in practice. The aim of this chapter is to contribute to that understanding. Team formation is investigated through the lens of Weick's "dropping the tools," which essentially implies that an organization or an individual has to set aside everyday practices (e.g., norms, values, rules, and routines) in order to survive (Weick, 1993). One such tool-dropping activity is forming new teams based on the situation, in much the same way as project managers do, for example, in product development (Engwall & Svensson, 2001, 2004), construction (Hällgren & Wilson, 2008), and engineering (Pavlak, 2004), as well as by leaders during firefighting (Bigley & Roberts, 2001). The chapter reports on a disaster on the shoulders of Mount Everest in 1996, but it extrapolates beyond the deaths of individuals and investigates team dynamics that are common in contemporary temporary organizations, albeit from a less traditional viewpoint. Thus, using the Mount Everest experience as a case, in a sense, heeds the call from Weick (2007) that academics should drop their tools too.

This chapter follows a practice approach (Schatzki, Knorr-Cetina, & Von Savigny, 2001; Whittington, 2006; Hällgren, 2009), meaning that extra attention was given to everyday activities, but with the disclaimer that in this case they had not been observed first hand but rather came from bibliographic accounts by the survivors. The chapter begins with a background on temporary organizations and how teams form within them. Then the chapter explains how dropping one's tools may be a necessity for overcoming problems, and it ends with a comparison between traditional temporary organizations (i.e., projects) and climbing expeditions as temporary organizations. Once the setting is established, the chapter revisits the Mount Everest events, which are later analyzed. Three different situated team types are suggested, initiated by five triggers. Finally, some implications are offered for academics, practitioners, and lecturers.

13.1. Background

13.1.1. Temporary Organizations

The concept and literature on temporary organizations originate from the 1960s, when Bennis (and colleagues) (1956, 1965, 1966) began talking about the need for *ad hoc*, temporary, groups in society. Later, Goodman and Goodman (1967, 1972, 1976) developed the idea, investigating theater plays. The reasoning was that a play was set up with a number of specialists who needed to be coordinated. Later, Lundin and Söderholm (1995) made further connections, not limited to projects, by publishing their account of the developments [see also Turner & Müller (2003)]. The main difference, compared to the permanent organization, as described by Cyert and March (1963), claim Lundin and Söderholm, is that it is assumed that a temporary organization is terminated at a certain point in time. An organization is terminated when the task is achieved, when the schedule ends, or when resources are exhausted. This would be equal to an end state (Lundin, 2009). In reality, some temporary organizations have a longer life cycle than a permanent organization. Nevertheless, an anticipated termination and end state causes a temporary organization to become action-oriented rather than decision-oriented.

A predetermined termination causes an organization to play by a different set of rules than if it were ongoing concern—different that is, from a permanent organization (Ekstedt, Lundin, Söderholm, & Wirdenius, 1999). It is commonly argued that the temporariness of temporary organizations makes planning a project easier than running operations in permanent organizations. The basis for these arguments is that fewer contextual influences interfere, and there is a limited goal to be achieved within a certain time. This has created a perception among the public that projects are rational (c.f. Hodgson, 2004; Cicmil, 2006). This sense of rationality is however sometimes hard to fathom when looking beneath the surface of a project, where it is evident that unexpected events (Söderholm, 2008), deviations (Hällgren & Maaninen-Olsson, 2005, 2009), changes (Dvir & Lechler, 2004; Steffens, Martinsuo, & Artto, 2007), crises (Loosemore, 1998; Gherardi, 2006), or risks (Akintoye & MacLeod, 1997; Baker, Ponniah, & Smith, 1999; Pender, 2001) are a part of everyday life in projects (Hällgren, 2009).

According to traditional ways of viewing project operations, such interruptions should be managed in a fairly standardized way (Packendorff, 1995; Cicmil & Hodgson, 2006; Blomquist, Hällgren, Milsson, & Söderholm, 2009). In reality, much of practice is very different and involves different means of achieving control. One such means is the creation of a response team.

13.1.2. Team Formation

Implementing teams of various kinds is frequently cited as useful (Bigley & Roberts, 2001; Engwall & Svensson, 2001, 2004; Pavlak, 2004; Hällgren & Wilson, 2008). A response team may go under such different names as task force, swat team, hot group, red team, tiger team, or cheetah team. These teams have the ability to question presupposed plans and routines that, together with task orientation, are the most common organizational governing mechanisms (Weick, 1976). The value of these kinds of teams essentially stems from "deep, intense, productive conflict—an open, honest struggle to reconcile opposing views" (Pavlak, 2004, p. 5) and are commonly associated with a high level of coordination and communication among their members (Pavlak, 2004).

While Pavlak focuses on the ability to question and generate new ideas, Engwall and Svensson (2001, 2004) are more concerned with practice as such. In an in-depth case study, they identified different types of teams dealing with problems in projects:

1. Project teams themselves
2. Tiger teams
3. Cheetah teams

These teams have different characteristics. They may:

1. Be explicitly sanctioned
2. Be mission-specific
3. Be nonpermanent
4. Have the full commitment of the members
5. Not be planned in advance

Project teams show the first three characteristics, tiger teams the first four, and cheetah teams all the features mentioned. Firefighting teams would fall within the tiger team features, as they are semipermanent teams sent out to respond to real fires. Engwall and Svensson called their particular formation "cheetah teams," referring to the big cat with incredible speed but no stamina. Hällgren and Wilson (2008), on the other hand, found that a dual structure was beneficial when attending to crises in construction projects. The response teams that were formed in their case were *ad hoc* formations based on predefined rules of the organization responding to a particular problem, thereby being similar, but not equal to, the cheetah team.

The previous information is relevant when considering team formation within the temporary organization. The authors all highlight the importance of understanding team formation and its characteristics in practice (Schatzki et al., 2001). There is, however, a lack of understanding of team formation in less traditional temporary organizations. It is believed that enhancing this understanding would also improve the understanding of how teams in general are formed.

13.1.3. Tool Dropping

Weick (1993) showed that in order to survive in some crisis situations, it may be absolutely essential to form a new team and take a step beyond the usual practices, which are essentially rules, norms, and values on which a practitioner draws. In a climbing project, for example, it is usual practice never to leave a fallen climber. In a more traditional project, such as product development or construction, usual practices include ways to apply risk management, file reports, make contact in case of emergency, and which regulations and laws to follow. [For further elaboration on the relationship among practice, practices, and practitioners, see Schatzki et al. (2001), Whittington (2006), or (Blomquist et al. (2009).] Some practices are internal and others are external to the practitioner; Jarzabkowski and Spree (2009) provide a review of the literature focusing on these aspects in the strategy-as-practice field. There is, however, common agreement on how decisions are made, and how they influence and are influenced by expectations about what should be done (i.e., practice) in a certain situation (Cyert & March, 1963, pp. 55–98). Actions according to expectations basically mean "given the situation, this will follow" (Olson, Roese, Zanna, Higgins, & Kruglanski, 1996; Zeelenberg, Van Dijk, Manstead, & Van Der Pligt, 2000) and thus it constitutes a planning mechanism for individual and organizational behavior—as so does practice.

Taking a step beyond obvious practices is similar to "dropping one's tools." Weick used the Mann Gulch disaster as a practical example of dropping one's tools. In that case, firefighters were sent to a bush fire and soon found themselves surrounded by fires. Despite trying to outrun the raging fire, some of them carried their heavy tools with them for a long time and for no purpose (Weick, 1993). In the case of Mann Gulch, Weick argued that disintegration of role structure and breakdown of sensemaking could have contributed to the disaster. The main problem, however, was that the identities of the firefighters were inscribed on their heavy equipment. As Weick (2007, p. 2) pointed out, Maclean (1992, p. 273) stressed the significance of the relationship between firefighters and their tools: "When a firefighter is told to drop his firefighting tools, he is told to forget he is a firefighter and run for his life." Furthermore, "When fire fighters are told to throw away their tools, they don't know who they are anymore, not even what gender" (Maclean, 1992, p. 226).

The tools in themselves represented an unwillingness to think anew and abandon what it meant to be a firefighter. Since then, "dropping the tools" has been used as a metaphor for making sense of a new environment and thinking about it in a different way (Weick, 1996; Jönsson, 2006; Weick, 2007). From here on in this chapter, the term *practices* is used interchangeably with the term *tools,* indicating that tools, like practices, are a governing mechanism of an organization.

13.2. Climbing Expeditions as Temporary Organizations

Climbing expeditions in the far Himalayas and temporary organizations have several features in common. Companies set up to guide clients to the top of Mount Everest even claim that project management is one of their core competencies (e.g., Adventureconsultants.co.nz. Accessed: 2008).

To start with the most obvious commonality, from the perspective of a project as a temporary organization, a climbing expedition is limited in time, scope, and resources and has a specific goal. Standing on the summit and making it back represents both the time and scope of an expedition. In a project, meeting time and scope requirements is often essential in the traditional viewpoint to delivering a successful project. Both expeditions and projects involve a team that is set up to achieve a task. In a commercial expedition, the team typically includes an expedition leader, a few guides, some Sherpas, and the clients. In addition, Sherpas are by definition subcontractors of the commercial expedition; subcontractors are also a feature found in software projects, where another company may do parts of a new development. In a traditional project, a team includes people with different areas of expertise. For example, in the construction of a diesel power plant, there may be a project manager and civil, electrical, and mechanical engineers. Commonly, both types of organization involve logistics: An expedition to the Himalayas requires arrangements for hundreds of kilos of resources such as oxygen, food, and water. In a road construction project, logistics may involve filling for the road and asphalt for surfacing it. Furthermore, in a project, risk management is a crucial part of the undertaking, just as it is on an expedition. While risk management in a project commonly involves some evasive strategies, it is much more physical on an expedition, where using harnesses, ropes, weather forecasts, and backup plans is the norm. Perhaps most important, it is common for external pressure on success to be great in both expeditions and projects; a commercial expedition must ensure future business and a project must succeed not only for the individual career but also because, in some cases, projects are make-or-break events for organizations. See Table 13.1 for a summary.

The comparisons reveal that mountain climbing expeditions do carry relevance for project teams (Kayes, 2004, p. 1282). Similarly, Hällgren (2007) proposed that mountain climbing expeditions carry extreme features of temporary organizations and thus provide lessons for projects, specifically on why and how actions come about. These arguments have validity and the approach based on

Table 13.1. Climbing Expeditions and Temporary Organizations

	Climbing Expeditions	Temporary Organizations
Time	Time objective	Time objective
Cost	Food, water, and gear	People, resources, etc.
Scope	Top (and back)	The project goal
Unique task	New situation each time	Often claimed feature of projects
Temporary endeavor	A noncontinuous climb within certain frames	A limited endeavor within the borders of a task
Team	The members of the expedition	The project team
Planning methods	The route and the necessary resources	The project plan and the resources to acquire it
Control methods	Physical securing during the climb and visible consolidation of personal and mountain conditions. Later on, the breakdown of the path into small objectives.	Project plan, work breakdown structure, earned value, CPM, PERT, etc.
Risk management	Mitigation of risks to circumvent them by, for example, choosing another path	The identification or risk to circumvent its impact
Management tasks	Logistics, stakeholders, subcontractors, clients, etc.	Logistics, stakeholders, subcontractors, clients, etc.

Source: Adapted from Hällgren (2007).

them is useful, especially in business environments increasingly organized into project teams and temporary groups with demanding goals and lean operations (Tempest, Starkey, & Ennew, 2007, p. 1044; Whittington, 1999).

13.3. Methodology

The qualitative research described in this chapter relies on one case, namely, the events that unfolded in 1996 on the shoulders of Sagarmatha, the Nepalese name for Mount Everest. The resemblance between temporary and climbing expeditions has already been demonstrated and emphasized here, and organizations by others (Elmes & Barry, 1999; Elmes & Frame, 2008; Hällgren, 2007, 2010; Kayes, 2002, 2004, 2005, 2006; Mangione & Nelson, 2003; Tempest et al., 2007). Following a similar path as previous researchers [Elmes & Barry,

1999; Kayes, 2002, 2004, 2006; Hällgren, 2007, 2010; Mangione & Nelson, 2003; Roberto, 2002; Rosen, 2007; Tempest et al., 2007; see Elmes & Frame (2008) for a notable exception], this chapter relies on personal accounts of events (Boukreev & DeWalt, 1998; Breashears, 1999; Gammelgaard, 1999; Krakauer, 1996, 1997; Kropp & Lagerkrantz, 2002; Weathers, 2000). Thus, the case is constructed around an understanding of the events through the texts of survivors. [See also Weick (1993) for a similar approach when investigating dropping the tools]. Although reconstructions of events and activities carry obvious limitations, those limitations also apply to other methods such as interviews (Orton & Weick, 1990). In one way the use of personal accounts is an allegory for academically dropping the tools and using alternative methods to understand a complex phenomenon (Weick, 1996, 2007). One inherent problem of qualitative in-depth studies is how to show the depth and amount of material while maintaining an entertaining story. Eisenhardt and Graebner (2007, p. 29) noted that "tables and other visual devices are central to signalling the depth and detail of empirical grounding." There are two tables in this chapter.

In terms of analysis, the accounts of the survivors were read and coded according to how the events unfolded. [The accuracy of the account based on the literature is further verified by Kayes (2004) and by Elmes and Barry (1999).] All in all, 26 activities or events were identified and put into a time line. There were 19 events associated to team formation (see Table 13.2). A team was coded as such when two or more people joined up to pursue a common goal (compare Engwall & Svensson, 2001, 2004), and then the team was coded as dropping the tools when there was no apparent connection to what would be expected in a given situation. For example, when expedition guides abandoned climbers, it was coded as dropping the tools; whereas tools were regarded as maintained when, for example, an expedition leader persevered with supporting the client. Analyzing the events by iterating between empirical data and emerging patterns (Langley, 1999), three situated patterns were identified,

1. Task
2. Survival
3. Rescue

and five triggers of team formation,

1. Rules
2. Goal achievement
3. Obstacle
4. Necessity
5. Expectations

The first three triggers are intimately associated with task orientation, the last two are associated with survival, and the first and last also with rescue orientation. The coding is reflected in Table 13.2.

13.4. The Mount Everest 1996 Disaster Revisited

There is a widely held view that the competition among the guiding expeditions contributed to the death of the individuals. Sir Edmund Hillary, among others, has criticized the developments that followed after wealthy businessman Dick Bass showed that any reasonably fit and wealthy person could get himself or herself to the top of Mount Everest by paying to do so. In fact, Hillary's criticism was seconded by one of the expedition leaders on Everest in 1996. To further emphasize the point, one of the clients in 1996 was Sandy Pittman, an extremely wealthy woman who took a considerable amount of luggage on her climb to the highest points of the seven continents. The availability process that started in the 1980s has not stopped since then, and there is plenty of evidence that it still exists (Elmes & Frame, 2008; Hällgren, 2010). This commercializations of Everest means that there are external forces exerting influence on expeditions. Since the expedition leaders are typically measured against their ability to put clients on the top of the mountain, their future revenues are dependent on reaching the peak successfully. This was true in 1996 and it has remained true years later. A second consequence of the development of commercial expeditions is that less experienced climbers find themselves stumbling around on the mountain. The implications are twofold:

1. There is an increased reliance on the leaders and guides, as fewer decisions can be taken by the individual; for example, it was even reported that some clients had never before tried climbing with crampons (Krakauer, 1997).
2. Individuals have less ability to make sound decisions on their own (such as whether to turn back), or for that matter, question decisions made by the expedition leader. Also, and possibly contributing to the disaster, Krakauer speculates that Hall had developed *hubris* after having been able to put a number of clients on the top and dealing satisfactorily with some hard situations on previous climbs.

The 1996 events started at the beginning of April, when the climbers started to gather in Kathmandu. Once the expeditions acclimatized to the lower parts of Everest and arrived at base camp, they discovered that several expeditions would be trying their luck on Everest. Among them was the Swede Göran

Table 13.2. Timeline and Team Formation Triggers

No.	Approx. time	Events/Activities	Description	Team Formation Trigger
1	< May 10	Delayed gear	Some essential gear was delayed coming in to Katmandu. Guides and clients stuck in Katmandu. Unplanned helicopter flight.	
2	< May 10	Base camp establishment	Due to bad weather, the establishment of the base camp was delayed by some days. Guides and clients stuck on the way.	
3	May 10	Broken promise	The Taiwanese expedition broke their promise not to start the ascent on May 9. Ended with three expeditions on the mountain simultaneously.	
4	5:40 am–7:10 am	Balcony	Krakauer et al. waited for expedition leader below the Balcony. The climbers waited for 90 minutes due to earlier decision.	Complying with rules, the climbers stopped, trying to reach the top.
5	~8 am	Climber short-roped	Sandy Pittman short-roped by ascending Sherpa.	Trying to make the client reach the goal, the top.
6	11 am–1 pm	Hillary's step	Krakauer et al. forced to wait below Hillary's step, as no ropes were fixed. Waited another hour for the ropes to be fixed.	Forced to wait because of obstacle, trying to reach the top
7	~1 pm	Queue	Waiting above Hillary's step allowed weaker climbers to form a queue.	Forced to wait because of obstacle, trying to reach the top.
8	~2:30 pm	Weather deterioration	2:30–3 pm it started snowing and a mist arrived.	
9	1:40 pm–3:10 pm	Summit wait	Guide waiting for clients more than 90 minutes, well after a safe turnaround time.	Complying with rules, the guide waited, trying to get people down safe.
10	3:30 pm	Expedition leader ascent	At 3.30 Fischer (expedition leader) was still ascending. Obviously slow and tormented.	
11	~3:40 pm	Pittman collapses	The previously short-roped climber (Pittman) collapses and has to be dragged down.	Expected to take care of fellow climbers, trying to get them down safe.
12	~4 pm	Fischer short-roped	Fischer hypothermic and seems to suffer from cerebral edema and is therefore short-roped.	Expected to take care of fellow climbers, trying to get them down safe.
13	4 pm	Climber waiting	Weathers' eyesight was lost on ascent. He promised to wait for descending guide. Ended up waiting for hours. Losing all eyesight. Gets help by Beidleman's (guide) ascending group.	Expected to take care of fellow climbers, trying to get them down safe.

14	4:31 pm	Hall and Hansen descend	In need of canned oxygen, Hall and Hansen are led to believe there are none.	Expected to take care of fellow climbers, trying to get them down safe.
15	5:30 pm	Weather deterioration	The weather changes to hurricane forces. A climber (Harris) walks off the edge and dies.	
16	8 pm	Huddle	Beidleman's group huddles in the hurricane as several are incapacitated and cannot locate themselves, and they otherwise run the risk of walking down the mountain.	Necessary in order to survive.
17	~8 pm	Limited backup	A guide (Boukreev) tries to summon people for a rescue attempt. He fails, as the others are worn out	
18	Night	Weather break	Beidleman locates some stars and some members of the team orient themselves to Camp IV, leaving five individuals behind.	Necessary in order to survive.
19	Night	Rescue attempt of huddle remains	Beidleman et al. meet Boukreev (guide), who saves 3 of the 5 individuals left behind.	Expected and part of the job to organize rescue attempt if fit, too dangerous to try high-altitude rescue.
20	Morning, May 11	Out of batteries	The disaster-struck expedition runs out of batteries. Another expedition refuses to give them batteries.	
21	9:30 am	Rescue attempt of Hall	Attempt to rescue Hall fails.	Expected and part of the job to organize rescue attempt if fit, too dangerous to continue.
22	~9:30 am	Rescue attempt of Fischer and Gau	Attempt to rescue Fischer and Gau is successful, but Fischer does not respond and is left for dead.	Expected and part of the job to organize rescue attempt if fit, too dangerous to try high-altitude rescue.
23	Morning	Search party for Weathers and Namba	A search party locates the remains of huddling group (Weathers and Namba). Alive but dying. Left to die.	Expected and part of the job to organize rescue attempt if fit, too dangerous to try high-altitude rescue.
24	4:30 pm	Climber returns from the dead	Weathers (client) staggers into camp. Is put in tent that blows to pieces overnight.	
25		Rescue preparations	Other expeditions on the mountain try to aid in the rescue of the exposed climbers.	Expected on moral grounds to aid rescue.
26	May 12	Helicopter rescue	The highest rescue ever performed helicopters Weathers and Gau off Everest.	Expected on moral grounds to aid rescue.

Kropp, who had ridden his bike from Sweden and would attempt a lone ascent. Kropp was first to start but turned back about a hundred meters from the top. "To turn around that close to the summit," Hall mused, shaking his head. "That showed incredibly good judgment on young Göran's part. I'm impressed" (Krakauer, 1996). A few days after what later unfolded as a disaster, Kropp successfully climbed Everest (Kropp & Lagerkrantz, 2002). Others ready to climb were Scott Fischer's team, Rob Hall's team, a South African team, a Taiwanese team, and an IMAX team with David Breashears. Most of these expeditions came to play a role in the unfolding of the later events.

In clear weather at midnight, May 9, 1996, three expeditions started for Camp IV to head to the summit. One of these teams, the one from Taiwan, broke an earlier agreement to limit the first push to two expeditions. This decision contributed to some later difficulties, as it added to the traffic jam on the way to the top. The other expeditions also had to take care of a sick climber who had been left to die. At midnight on May 9, the three expeditions broke out from Camp IV and headed for the summit. Some of the climbers were much stronger and were soon ahead of the others. However, when Krakauer and part of Hall's team arrived at the Balcony, they waited as ordered in the blisteringly thin air. During the more than 90 minutes they spent waiting, several climbers passed them. At 7:10 am, Krakauer was able to continue, and as he passed, he noticed a Sherpa short-roping one of the clients (i.e., tying a short rope between them, literally dragging the other climber up the mountain). Another issue that has caused heated debate between Krakauer and Boukreev is whether it was a good idea for Boukreev to ascend without oxygen. Krakauer (1996, 1997) expressed doubts, while Boukreev (Boukreev & Dewalt, 1998) argued that he was used to going without supplemental oxygen and that he probably would be of less help if he used it then. The fact remains that, as an individual, Boukreev made two solo rescue attempts during hurricane conditions, saving several climbers.

A notorious bottleneck, Hillary's Step, is a very hard passage. To get around it there had been an agreement that Sherpas would climb in advance and fasten ropes. When Krakauer arrived at 11 am, no ropes were to be found and they were forced to stop for about an hour. Finally, when no one put up the ropes, the climbers who had arrived first at the scene decided to put up the ropes themselves, which took about an additional hour. By then the climbers were scattered over the area above Camp IV (at about 8000 meters of the 8848 meters to the summit). Meanwhile a queue was building up. This would not have been too much of a problem had not communication options been limited to a few radio devices carried by only a small number of the guides. The communication limitations, in combination with all decisions made on the mountain having to be made by the expedition leader, contributed to a serious decision problem when the situation became precarious. At 11:30 am three climbers decided to turn back, just before they attempted to climb Hillary's Step.

Climbing Everest is not easy, and it takes a lot of time. That is why the expeditions departed around midnight. To be safe, climbers have to turn back from the summit around 1 pm the following day; returning at 2 pm is risky, and a return at 3 pm places climbers in serious danger. The first climbers made it to the top around noon, which was less risky. More worrying was that the last climber to reach the summit turned around at 3:30 pm, long after the safe hour to return. When the first ascenders came back to the notorious Hillary's Step, Krakauer was alarmed: "Thirty feet below, some 20 people were queued up at the base of the Step, and three climbers were hauling themselves up the rope that I was attempting to descend. I had no choice but to unclip from the line and step aside." Krakauer ended up waiting for about an hour, while his supplemental oxygen was running out. Soon he was incorrectly advised that there was no more supplemental oxygen in the stash below the Step. Upon closer examination, some bottles proved to be full, but one of the guides, suffering from hypoxia, refused to believe it.

By 2:30 to 3:00 pm, it started to snow and mist arrived. Soon the weather grew worse, and the wind became a blizzard with hurricane forces that blocked vision and increased wind chill. The problems for the climbers who were scattered on the shoulders of the mountain had only just begun. One of the guides, Beidleman, waited more than 90 minutes on the summit, until 3:10 pm, for the last client, Gammelgaard, to arrive (Gammelgaard, 1999). By then Beidleman was worried, especially because Scott Fischer, his expedition leader, had not yet arrived. The group passed Fischer, still ascending, about 20 minutes from the top. Soon Pittman, the short-roped climber, collapsed and had to be dragged down the mountain. Fischer eventually made it to the top at 3:30 pm, about the same time as his client Hansen. Expedition leader Hall and Hansen then descended. By that time Fischer was showing signs of cerebral edema and hypothermia, which made his behavior confused and irrational. He was eventually short-roped by one of the Sherpas. The Sherpa, however, had to leave him behind around 10 pm, together with Makalu Gau (another expedition leader) and three other Sherpas who had arrived with the latter. They soon descended to get help.

At 4 pm Krakauer met Beck Weathers, who wrote his own account (Weathers 2000), a fit climber who had stopped after experiencing eye problems. Weathers had promised his expedition leader that he would wait and was now very cold. Even though Krakauer urged him to accompany them, he refused and waited for a guide to help him down. This decision would later bring him as close to death as anyone can come and cost him the loss of fingers, toes, his nose, and ears. When the Beidleman group caught up with Weathers, they helped him and Namba (yet another climber) down the slopes. At 5:30 pm Krakauer was very close to Camp IV, but the wind had also picked up to hurricane level. Afraid of making a mistake in the poor visibility, Krakauer sat down to prepare himself for a short, 200-yard descent. As he puts it: "There were zero margins for error. Worried about making a critical blunder, I sat down to marshal my energy" (Krakauer, 1996). Harris, one of the guides, passed him. While Krakauer believed that Harris had made it back, eventually he was found to have walked off the cliff and died. At roughly 4:30 pm, Anatoli Boukreev, one of the guides, made it back to Camp IV (Boukreev & DeWalt, 1998). At 4:31 pm Hall reported that he and Hansen were stuck above Hillary's Step and were in great need of supplemental oxygen. Even though there were actually two full bottles waiting a few hundred yards below, they had been reported empty by others farther down the mountain.

At 8 pm the Beidleman group, which consisted of guides Beidleman and Groom, two Sherpas, and seven clients, was left in the pitch-black evening on Everest with hurricane forces blowing. Several of the climbers were more or less incapacitated and could make very little effort to walk. One of the climbers who had previously made it to Camp IV tried to rouse other worn-out climbers but did not succeed. Six times he tried to locate missing climbers but always came back empty-handed.

During the night, Beidleman was able to locate some stars through the wind drifts and from them he could get a sense of their precarious location on the south col of Mount Everest. Clients Fox, Namba, Pittman, and Weathers were incapacitated and could not walk. Client Madsen volunteered to wait while the others made a push for Camp IV. On the way to the camp they met Boukreev, who went looking for the missing climbers and eventually was able to locate the huddling climbers and save Fox, Pittman, and Madsen. He left Namba and Weathers behind because he thought that they were dead. Hall made a series of transmissions during the night but seemed to have become increasingly ill and disillusioned. At 9:30 pm two Sherpas tried to make it back and save Hall. They did not succeed. Four other Sherpas tried to reach Fischer and Gau. Fischer did not respond, so they left him and made it back with Gau. At 6:20 am a radio call was patched through to Hall's pregnant wife in New Zealand. Hall ended the conversation with an attempt to comfort her: "I love you. Sleep well, my sweetheart. Please don't worry too much" (Krakauer, 1996).

The next morning, May 11, the surviving climbers learned that Namba, Weathers, and Fischer were still stuck on the mountain. By then the batteries for the radios had failed. One of the climbers asked another expedition for some batteries but was refused. Another search party was sent out, and they were easily able to locate Namba and Weathers, who were both breathing although their faces were covered by seven centimeters of ice. Rather than risking the entire group, they decided to leave the climbers behind. The search party made it back at 8:30 am. At 4:30 pm, Weathers staggered back to Camp IV. Weathers was put in a tent that collapsed overnight. He had no way of communicating with others. Once again he was brought to the brink of death (Weathers, 2000). Meanwhile, the IMAX team and other teams at base camp started to climb toward higher camps in order to aid the rescue of the remains of the expeditions.

On May 12, Weathers and Gau were brought down to Camp II, where the highest rescue attempt ever tried was conducted. Weathers and Gau were later flown to hospitals for treatment of severe frostbite. At the end of the events several people had lost their lives, even more were physically and/or mentally hurt, and eventually the events would set a footprint in the annals of mountaineering as one of the worst accidents ever. Of course there are multiple explanations as to complex events, but just blaming the weather is to simplify the events. As Krakauer laments, which is relevant to the literature on temporary organizations, at the end "the clock had as much to do with the tragedy as the weather, and ignoring the clock can't be passed off as an act of God" (Krakauer, 1996).

The observations and the case are built around the reproduction of the events as described by the survivors' accounts (Krakauer, 1996, 1997; Boukreev & DeWalt, 1998; Breashears, 1999; Gammelgaard, 1999; Weathers, 2000; Kropp & Lagerkrantz, 2002). The result and a timeline over the events and activities are reflected in Table 13.2.

13.5. Discussion

The aim of this chapter is to contribute to the understanding of situated teams as a common practice in project organizations. Examining the events that killed nine people, this chapter takes the reader beyond the death of individuals and investigates the features of response teams common in most of today's organizations (Tempest et al., 2007; Whittington et al., 1999). The chapter thus adds to the theoretical and practical understanding of contemporary temporary organizational forms in general and situated team formation in particular. The analysis relies on the events on Mount Everest in 1996 and several accounts thereof (Krakauer, 1996, 1997; Boukreev & DeWalt, 1998; Breashears, 1999; Gammelgaard, 1999; Weathers, 2000; Kropp & Lagerkrantz, 2002).

A new cheetah or response team within a current team (i.e., a project) is a form of situated team and a matter of "deep, intense, productive conflict" (Pavlak, 2004, p. 5) that becomes "decoupled from other issues on the project agenda" (Engwall & Svensson, 2004, p. 299). Therefore, there are questions and views of the world from a different perspective, and these suggest a change in practices, rules, traditions, or norms (Whittington, 2006), aare more or less formal or informal, and are more about abstract expectancies concerning what should be done (Olson et al., 1996).

In some situations, sticking to practices is beneficial and at other times directly harmful. In this regard, Weick (1993) has shown the importance of "dropping the tools" (i.e., the practices) in order to think differently about a crisis situation. Engwall and Svensson (2004) identified three general types of cheetah teams: focus, rescue, and a mix thereof. Personnel in a focus team already belong to a project when disaster strikes, and the formation of the focus team sets their attention in a specific direction. The rescue team, on the other hand, consists of personnel from outside the project focusing on the problem at hand. The third alternative, the mix, has both external and internal personnel. These teams are all explicitly sanctioned, have a specific mission, are not permanent, require full membership, and are not planned in advance. In addition, Engwall and Svensson (2004) found that cheetah teams were triggered by sudden or emergent emergencies. Although these findings are important and add to an understanding of the inner workings of a project, they do not elaborate on what triggers cheetah teams. By iterating between data and theory until patterns emerged (Langley, 1999), three more general types of situated teams were identified—task, survival, and rescue—as well as five more detailed triggers—rules, goal achievement, obstacle, necessity, and expectations. Table 13.2 specifies which triggers belong to each event/activity. This and following analysis should convey the "rigor, creativity and open-mindedness of the research process" (Eisenhardt & Graebner, 2007, p. 30). Two points can initially be made. First, one could argue that the each type of team is a process model of team formation. However, in the Mount Everest case the teams were not consistent over time, breaking up and coming together continuously in new formations, and each team formation had its own content and features. Therefore, it is three different types rather than processes that are discussed. Second, the triggers are slightly more detailed than those in Engwall and Svensson (2004). Their triggers are still valid, but they are less detailed and explain less of the intricacies of team formation.

Having illustrated that an expedition is a temporary organization sharing several features with more traditional projects [see Table 13.1, Hällgren (2007, 2010), and Kayes (2004)], the following sections elaborate on the specificities of the teams and transfer insights gained from expeditions to more traditional projects.

13.5.1. The Task-Situated Team

Task teams follow one of the two most common governing mechanisms in organizational behavior, the task (Weick, 1976). A task-oriented team is formed because there is a task that needs to be achieved. The reason it needs to be achieved varies, but in the case of Mount Everest, the obvious task, or mission, is to make it to the summit and back. Although it is important to survive, one could argue that reaching the peak of Everest was the looming goal for several of the climbers and that surviving was the subordinate goal. This is evidenced in how guides short-roped climbers in order to put them on the top (5*).

Although assembled around a task (e.g., the "bagging" of the top in the Everest case, or achievement of the project goal and scope in a construction project), a task team is triggered by different things. In the Mount Everest case, three general triggers were associated with the task team. *Rule-triggered teams* are formed because the rules say so; for example, Krakauer and his team waited on the Balcony as determined in advance (4). These kinds of triggers are thus set in place by predefining milestones or gates and are closely associated with the general task that needs to be done. This is worth noting because typically a task and rules are seen as two separate entities. Here they are both necessary to fulfill a mission while having a team-forming function. *Goal achievement–triggered teams* are formed naturally enough, in order to reach the goal. In these situations the goal overshadows most other things, including threats to life [e.g., short-roping Pittman toward the top and thus using valuable energy (5)]. These teams can be found, for example, in skunk work projects, which can of course be very valuable (Christensen & Kreiner, 1991; Lindahl & Rehn, 2007) but also very harmful to the organization. *Obstacle-triggered situated teams* were formed whenever there was a physical non-milestone obstacle that forced one or several persons to wait until the path was cleared, which was the case below Hillary's Step (6) and the queue on the way down (7). In a traditional setting this could be a group of people trying to solve a technical problem in a construction project (Hällgren & Wilson, 2008) or a product development project (Engwall & Svensson, 2001, 2004).

A task team resembles a focus team in that it has personnel who are already present and contribute to a focus. However, a task team is formed before a disaster strikes. The feature that the team is formed before a disaster indicates that the team is formed in compliance with practices. Practices state what should be done in a certain situation, and the team is formed because practices are adhered to without finding an alternative route (Weick, 1993). There is an inherent danger in a task team because its activities are geared toward progress rather than evaluating the situation at hand. The practices that are available to a person tend to make this person behave as expected and not question whether the practices could be problematic. In an accumulating crisis, taking this type of path seems to be risky, as it leads deeper into the crisis.

---

* Numbers in parentheses correspond to the numbers in Table 13.2.

13.5.2. The Survival-Oriented Situated Team

Survival teams are concerned with one of the most basic human behaviors: survival. Here practices are set aside as lives are threatened. Weicks (1993) described firefighters who viewed the situation in a completely different way from their unfortunate colleagues and survived. In that case, they had burned a wide circle and laid down in the ashes, so that when the fire arrived, it roared around rather than over them. On Mount Everest, it meant, for example, abandoning fallen team members (18), which in one way goes against what a moral person is expected to do at sea level, but which in the extreme altitude of Mount Everest is sometimes necessary for survival. In other cases it meant adhering to the same practices, such as when Pittman collapsed (11).

The triggers for survival teams are different from those for task teams. The former are triggered by necessity and expectations once a disaster is present. *Necessity-triggered teams* are formed from a collective insight that a group of individuals needs to stick together to stay alive. The individuals find comfort in each other, while their joint efforts may lead to a better solution or taking care of one other. Hence, the team members are pushed together by an event that draws their attention. In the Mount Everest case the event was mostly external [e.g., a climber waiting (13), huddle (16), and a break in the bad weather (18)], but in another setting a situation may well be internal. This trigger varies between dropping previous practices (18) or not (13, 16); when it becomes necessary to take an alternative route, people seem to do that. *Expectations-triggered teams* form when people are supposed to take care of each other despite risk to their own life, such as when Pittman collapsed (11), when Fischer was short-roped (12), and when Hall and Hansen descended (14). When triggered by this mechanism, people seem to be comforted not by another's presence but rather by knowing they are doing the "right thing" or what is expected from them. In the Mount Everest case, the expectations-triggered teams seem to have been the most dangerous, because the people in these teams were in a very dangerous situation and became enmeshed in practices that they had been taught and had used previously. These practices hindered them from making a sound judgment about what could have been a good decision from a personal point of view. In a traditional project, a comparable situation could be when a project manager confuses certain risk management practices with an appropriate response to a much more difficult situation (Pender, 2001).

Survival teams are a consequence of an accumulative or a sudden crisis and form only once a crisis is realized. The team formation therefore has one fundamental feature: It results from a collective insight by the members of the organization. In the Mann Gulch case the group realized this collective insight and broke into several more or less independent teams without united leadership

(Weick, 1993). However, in contrast to Mann Gulch, on Mount Everest the breakdown of leadership contributed to some members staying alive. Both expedition leaders Hall and Fischer died, and the third leader, Gau, was severely injured. They lacked communication channels and left their expedition members to fight on their own. If the climbers had relied on formal decisions and their leaders, more climbers would presumably have died. This notion, that a breakdown of leadership saved lives on Mount Everest, is contrary to popular belief [e.g., Krakauer (1997) and Mangione & Nelson (2003)].

13.5.3. The Situated Team Oriented toward Rescue

Rescue teams, as shown by Engwall and Svensson (2004), use personnel who are located outside the immediate events and help in a crisis. Rescue teams, however, also are unique in the sense that these teams are governed not by explicit sanctions but rather by what is expected of the individual or organization, as was the case, for example, of the rescue attempt of the huddle (19). This type of team relies on tools that are taught and practiced, as well as expectations, that is, what a person is expected to do from a moral standpoint (Olson et al., 1996). By its very nature, rescue is an unselfish act because one does it for someone else and sometimes at risk to one's own life. When Boukreev went out in the storm by himself to find lost climbers is an example (17).

The trigger of a rescue team is associated with expectations and rules. People undertake rescues either because it is the organization's rule that it is their task to come to someone's aid, or because it is expected of them. The first type is a rule-triggered rescue team. Fire brigades, established in advance to aid in case of a fire, are a good example. This type of deployment (or trigger) has been shown to increase reliance on such teams (Bigley & Roberts, 2001) and assumes that these teams rely heavily on the practices of their organization and playing by the book. Examples from the Mount Everest case are the rescue attempt of the huddle (19), the rescue attempt by Hall (21), the rescue attempt by Gau and Fischer (22), and the search party (23). In all these situations, the guides and Sherpas performed their predefined roles in spite of the threat to their own lives. In some cases they did not get help from other guides or Sherpas, and this suggests that one needs to buy into the rules in order to be part of the team. This trigger applies to people within an organization rather than people external to it.

If people are external to an organization they have not accepted the rules beforehand and therefore play by logic, and expectations. *Expectations-triggered rescue teams* are teams formed by the mindset that "given this situation, I ought to do this." Hence they are triggered by expectations (Olson et al., 1996) (Zeelenberg et al., 2000), and expectations act as a planning mechanism of their own. The IMAX expedition, for example, did not have any obligation to come to the aid of the troubled expeditions, and doing so in fact endangered their own expedition because the rescue used up valuable energy and supplies (25, 26). Because of moral obligations, they did help by any means they found possible. This is in contrast to another expedition that refused to lend batteries needed for communication when the disaster was an established fact (26). This type of trigger allows teams to define the means of action on their own. It is closer to an acceptance of how something ought to be done rather than a predefined action pattern that determines what should happen. This type of trigger thus assumes more direct coordination because there are fewer predefined rules to govern the behavior of the individuals on the teams.

Similar to a survival team, a rescue team is closely associated with a declared crisis. It does not seem to matter whether the crisis occurs over time or suddenly; rather, it matters that it is an acknowledged emergency. In contrast to survival teams but similar to task teams, rescue teams adhere to established practices that allow the teams to be initiated and placed into action. They resemble tiger teams in that they play by a set of rules and therefore, in some ways, are defined beforehand (Pavlak, 2004). This type of team further resembles focus teams and the rescue teams described by Engwall and Svensson (2004). Rule-triggered rescue teams are similar to focus teams because personnel come from within the organization, and expectation-triggered rescue teams are similar to rescue teams whose personnel are external to an organization. The mixed teams identified by Engwall and Svensson may therefore have different triggers and may not be a third type of team.

13.5.4. Toward an Increased Understanding of Situated Teams

Although appreciated in practice, the deployment of situated teams is neglected by academia in general and project management in particular. It is seldom touched upon in theory. Following Engwall and Svensson's (2004) definition of team types, task, survival, and rescue teams are slightly different. They do share characteristics that include having a specific mission, being temporary, having full commitment from their members, and being supposed to dissolve. However, *they are not formally sanctioned*. The teams that were identified can be seen in three different relations to the crisis at hand. A task team is functioning while the task is still valid and the crisis has not yet got the situation in its grip. At this juncture, there is still no common consensus as to whether there is a crisis. A survival team, on the other hand, is caught in the crisis and the team forms as the situation progresses. A rescue team, by contrast, is mainly responsive and tries to resolve a problematic situation. Significant in all these types of teams is that there is no central authority deciding who should team up with whom, which rather is a result of situational needs. This leaves the identified team formations somewhere between that of a tiger team (Pavlak, 2004) and a cheetah team (Engwall & Svensson, 2001, 2004), and far from being a project team (Lundin & Söderholm, 1995). More important, teams do not form (only) because of a formal decision. They emerge naturally depending on the situation at hand. It seems to be a basic organizational principle for which this case provides evidence. The triggers noted in previous research (Engwall & Svensson, 2001, 2004) seem to be on a too high a level to explain the intricacies of a situation. More attention to detailed accounts indicates other triggers explain why different types of teams occur, including cheetah teams. These triggers are closely linked to a crisis, as noted previously. Some triggers require a team as a whole to drop the tools on which it had previously relied to function in its organization and society. This dropping allows a team to redefine the purpose of its existence, and quite often another team appears to be born out of it. As can be seen in the Mount Everest case, dropping the tools may in some situations make the difference between life and death.

13.6. Implications

Engwall and Svensson (2001, 2004) call cheetah teams the most flexible of project team remedies to resolve unexpected events. Essentially, they form as a crisis is identified. However, cheetahs may be the fastest animal on the earth, but if a cheetah team requires formal sanctioning it may deploy too late to remedy a crisis. This suggests that there is more to cheetah teams than can be discovered through traditional project research. This strengthens the argument that less traditional industries need to be researched to increase knowledge about temporary organizations and projects. However, this chapter relies on only one case in a less traditional setting. This implies there is also a need to elaborate on findings from more traditional industries. Another possible avenue for further research is to examine the various triggers identified in this case study as well as what can be identified from other cases, to see how they influence team formation. Detailed accounts of events seem to be useful and important in developing such understanding. Lastly, since these teams are formed out of unintended events, research could pay more attention both to intended team formations and unintended team formations. In the case of the latter, the unexpected event becomes the trigger and needs extra scrutiny.

The practical implications of this chapter are in part obvious. Any organization wishing to be able to cope with the unexpected needs to be able to form new teams depending on what a situation requires. Some of these teams may be intentional, such as the formal sanctioning of cheetah teams, but others may be unintentional and form out of a situational necessity. Either way, the study of team formation suggests that project expertise and knowledge involve far more than the application of plans and methods (e.g., Dvir & Lechler, 2004; Hällgren & Maaninen-Olsson, 2009). The ability to form effective teams in certain situations requires rethinking and questioning, capabilities that any project manager should value. Something very seldom addressed by this type of investigation is education. When academics stick to traditional industries, they may limit their research and the resulting understanding. Acknowledging less traditional industries may enable them to "tell a good story" that catches much attention. Because of that attention, learning from such a story could be lasting. Not only practitioners and academics should drop their tools, but also academics in their research, and when teaching.

13.7. Conclusion

The aim of this chapter was to increase understanding of a practical phenomenon. Three types of teams were identified: task, survival, and rescue. Similar to cheetah teams (Engwall & Svensson, 2004), these teams are not planned in advance when the project is set up, but, in contrast, upper management formally sanctions cheetah teams. They occur in different phases of a crisis and are initiated by five different triggers: rules, goal achievement, obstacle, necessity, and expectations. Previous analysis has not paid attention to the phase in which the team is formed and thus excluded any explanation of the various triggers that were released. Furthermore, contrary to approaches common in construction, engineering, and product development, this chapter paid attention to a less traditional industry, climbing expeditions, and thereby incorporated additional elements to the explanations of team formation, such as the survival of individuals rather than organizations and projects.

Consequently, this chapter elaborated on the notion of the temporary organization (Bennis, 1965; Lundin & Söderholm, 1995) because it suggests that there are additional forms of temporary organization. That is, whenever a team forms, organizations emerge that are even more temporary than in a standard project, and they play by a different set of rules from those applied in other temporary organizations. This has already been noted by Engwall and Svensson, but the observations in this chapter depart from theirs by suggesting that which type of team is formed depends on a phase and a trigger. The main message of this chapter is that, in order to survive and thrive, it may become necessary to question even those practices that are most deeply ingrained and taken for granted. Such questioning could alter the entire logic on which most projects rely.

Acknowledgments

I am sincerely grateful for the support of this project provided by Ragnar Söderbergs foundation, and the comments provided by the participants at Colloque en gestion de projet in Montreal.

References

Akintoye, A. S., & Macleod, M. J. (1997). Risk analysis and management in construction. *International Journal of Project Management, 15*(1), 31–38.

Baker, S., Ponniah, D., & Smith, S. (1999). Risk response techniques employed currently for major projects. *Construction Management & Economics, 17*(2), 205–213.

Bennis, W. G. (1965). Beyond bureaucracy. *Trans-Actions,* July–August, 31–35.

Bennis, W. G. (1966). Organizational revitalization. *California Management Review, 9*(1), 51–61.

Bennis, W. G., & Shepard, H. A. (1956). A Theory of Group Development. *Human Relations, 9*(4), 415–437.

Bigley, G. A., & Roberts, K. H. (2001). The incident command system: High reliability organizing for complex and volatile task environments. *Academy of Management Journal, 44*(6), 1281–1299.

Blomquist, T., Hällgren, M., Nilsson, A., & Söderholm, A. (2009). Projects-as-Practice: Making project research matter. *Project Management Research, 41*(1), 5-16.

Boukreev, A., & Dewalt, G. W. (1998). *The Climb: Tragic Ambitions on Everest.* New York: St. Martin's Press.

Breashears, D. (1999). *High Exposure: An Enduring Passion for Everest and Unforgiving Places.* New York: Simon & Schuster.

Christensen, S., & Kreiner, C. (1991). *Projektledning, att leda och lära i en ofullkomlig värld.* Lund, Sweden: Academia Adacta.

Cicmil, S. (2006). Understanding project management practice through interpretive and critical research perspectives. *Project Management Journal, 37*(2), 27–37.

Cicmil, S., & Hodgson, D. (2006). Making projects critical: An introduction. In S. Cicmil, & D. Hodgson (Eds.), *Making projects critical.* New York: Palgrave Macmillan.

Cyert, R. M., & March, J. G. (1963). *A behavioral theory of the firm.* Englewood Cliffs, NJ: Prentice-Hall.

Dvir, D., & Lechler, T. (2004). Plans are nothing, changing plans is everything: The impact of changes on project success. *Research Policy, 33*(1), 1–15.

Eisenhardt, K. M., & Graebner, M. E. (2007). Theory building from cases: Opportunities and challenges. *Academy of Management Journal (AMJ), 50*(1), 25–32.

Ekstedt, E., Lundin, R. A., Söderholm, A., & Wirdenius, H. (1999). *Neo-industrial organizing: Renewal by action and knowledge formation in a project-intensive economy.* London: Routledge.

Elmes, M., & Barry, D. (1999). Deliverance, denial, and the death zone. *Journal of Applied Behavioral Science, 35*(2), 163–187.

Elmes, M., & Frame, B. (2008). Into hot air: A critical perspective on Everest. *Human Relations, 61*(2), 213.

Engwall, M., & Svensson, C. (2001). Cheetah teams. *Harvard Business Review, 79*(2), 20–21.

Engwall, M., & Svensson, C. (2004). Cheetah teams in product development: The most extreme form of temporary organization? *Scandinavian Journal of Management, 20*(3), 297–317.

Gammelgaard, L. (1999). *Climbing high: A woman's account of surviving the Everest tragedy.* Seattle, WA: Seal Press.

Gherardi, S. (2006). *Organizational knowledge: The texture of workplace learning.* Oxford, UK: Blackwell.

Goodman, L. P., & Goodman, R. A. (1972). Theater as temporary system. *California Management Review, 15*(2), 103–108.

Goodman, R. A. (1967). Ambiguous authority definition in project management. *Academy of Management Journal, 10*(4).

Goodman, R. A., & Goodman, L. P. (1976). Some management issues in temporary systems: A study of professional development and manpower—The theater case. *Administrative Science Quarterly, 21*(September), 494–501.

Hällgren, M. (2007). Beyond the point of no return: On the management of deviations. *International Journal of Project Management, 25*(8), 773–780.

Hällgren, M. (2009). *Avvikelsens mekanismer: Observationer av projekt i praktiken [The mechanisms of deviations: Observations of projects in practice].* Doctor, Umeå University, Umeå.

Hällgren, M. (2010). Groupthink in temporary organizations. *International Journal of Managing Projects in Business, 3*(1), 94-110.

Hällgren, M., & Maaninen-Olsson, E. (2005). Deviations, uncertainty and ambiguity in a project intensive organization. *Project Management Journal, 36*(1), 17–26.

Hällgren, M., & Maaninen-Olsson, E. (2009). Deviations and the breakdown of project management principles. *International Journal of Managing Projects in Business, 2*(1), 53–69.

Hällgren, M., & Wilson, T. (2008). The nature and management of crises in construction projects: Projects-as-practice observations. *International Journal of Project Management, 26*(8), 830–838.

Hodgson, D. E. (2004). Project work: The legacy of bureaucratic control in the post-bureaucratic organization. *Organization, 11*(1), 81–100.

Jarzabkowski, P., & Spee, A. P. (2009). Strategy-as-practice: A review and future directions for the field. *International Journal of Management Reviews, 11*(1), 69–95.

Jönsson, S. (2006). On academic writing. *European Business Review, 18*(6), 479–490.

Kayes, C. D. (2002). Dilemma at 29000 feet: An exercise in ethical decision making based on the 1996 Mt. Everest climbing disaster. *Journal of Management Education, 26*(3), 307–321.

Kayes, C. D. (2004). The 1996 Mount Everest climbing disaster: The breakdown of learning in teams. *Human Relations, 57*(10), 1263–1284.

Kayes, C. D. (2005). The destructive pursuit of idealized goals. *Organizational Dynamics, 34*(4), 391–401.

Kayes, C. D. (2006). *Destructive goal pursuit: The Mount Everest disaster.* New York: Palgrave Macmillan.

Krakauer, J. (1996) http://outside.away.com/outside/destinations/199609/199609_into_thin_air_1.html, accessed 090909.

Krakauer, J. (1997). *Into thin air.* London: PanMacmillan.

Kropp, G., & Lagerkrantz, D. (2002). *Göran Kropp.* Stockholm: Liber.

Langley, A. (1999). Strategies for theorizing from process data. *Academy of Management Review, 24*(4), 691–710.

Lindahl, M., & Rehn, A. (2007). Towards a theory of project failure. *International Journal of Management Concepts and Philosophy, 2*(3), 246–254.

Loosemore, M. (1998). Organisational behaviour during a construction crisis. *International Journal of Project Management, 16*(2), 115.

Lundin, R. (2009). End states and temporary organizations. *Euram,* Liverpool, UK.

Lundin, R. A., & Söderholm, A. (1995). A theory of the temporary organization. *Scandinavian Journal of Management, 11*(4), 437–455.

Maclean, N. (1992). *Young men and fire.* Chicago: University of Chicago.

Mangione, L., & Nelson, D. (2003). The 1996 Mount Everest tragedy: Contemplation on group process and group dynamics. *International Journal of Group Psychotherapy, 53*(3), 353–373.

Olson, J. M., Roese, N. J., Zanna, M. P., Higgins, E. T., & Kruglanski, A. W. (1996). Expectancies. In A. W. Kruglanski & E. T. Higgins (Eds.), *Social psychology: Handbook of basic principles.* New York: Guilford Press.

Orton, D. J., & Weick, K. E. (1990). Loosely coupled systems: A reconceptualization. *Academy of Management Review, 15*(2), 203–223.

Packendorff, J. (1995). Inquiring into the temporary organization: New directions for project management research. *Scandinavian Journal of Management, 11*(4), 319–333.

Pavlak, A. (2004). Project troubleshooting: Tiger teams for reactive risk management. *Project Management Journal, 35*(4), 5–14.

Pender, S. (2001). Managing incomplete knowledge: Why risk management is not sufficient. *International Journal of Project Management, 19*(2), 79–87.

Roberto, M. A. (2002). Lessons from Everest: The interaction of cognitive bias, psychological safety, and system complexity. *California Management Review, 45*(1), 136–158.

Rosen, E. (2007). Somalis don't climb mountains: The commercialization of Mount Everest. *Journal of Popular Culture, 40*(1), 147–168.

Schatzki, T. R., Knorr-Cetina, K., & Von Savigny, E. (2001). *The practice turn in contemporary theory.* New York: Routledge.

Söderholm, A. (2008). Project management of unexpected events. *International Journal of Project Management, 26*(1), 80–86.

Steffens, W., Martinsuo, M., & Artto, K. (2007). Change decisions in product development projects. *International Journal of Project Management, 25*(7), 702–713.

Tempest, S., Starkey, K., & Ennew, C. (2007). In the Death Zone: A study of limits in the 1996 Mount Everest disaster. *Human Relations, 60*(7), 1039–1064.

Turner, R. J., & Müller, R. (2003). On the nature of projects as a temporary organization. *International Journal of Project Management, 21*(1), 1–8.

Weathers, B. (2000). *Left for dead: My journey home from Everest.* New York: Villard.

Weick, K. E. (1976). Educational organizations as loosely coupled systems. *Administrative Science Quarterly, 21*(1), 1–19.

Weick, K. E. (1993). The collapse of sensemaking in organizations: The Mann Gulch disaster. *Administrative Science Quarterly, 38*(4), 628–652.

Weick, K. E. (1996). Drop your tools: An allegory for organizational studies. *Administrative Science Quarterly, 41*(2), 301–313.

Weick, K. E. (2007). Drop your tools: On reconfiguring management education. *Journal of Management Education, 31*(1), 5–16.

Weick, K. E., & Sutcliffe, K. M. (2001). *Managing the unexpected.* San Francisco: Jossey-Bass.

Whittington, R., Pettigrew, A., Peck, S., Fenton, E., & Conyon, M. (1999). Change and complementarities in the new competitive landscape: A European panel study, 1992–1996. *Organization Science, 10*(5), 583-600.

Whittington, R. (2006). Completing the practice turn in strategy research. *Organization Studies, 27*(5), 613–634.

Zeelenberg, M., Van Dijk, W. W., Manstead, A. S. R., & Van Der Pligt, J. (2000). On bad decisions and disconfirmed expectancies: The psychology of regret and disappointment. *Cognition & Emotion, 14*(4), 521–541.

Part Three

Lessons to Be Learned

The following five chapters show clearly that an expedition is an active project and, above all, that genuine tension exists between the need to anticipate, prepare, and plan on the one hand and the need to remain flexible on the other.

The first part of the solution to this tension requires that everyone accept the two components of the rigidity–flexibility paradox. Their coexistence is necessary. It seems impossible to dispense with either one. The second part of the solution is to strike a balance or, better still, take up a position on a continuum that may not always constitute the point of balance. Furthermore, tension can be expressed more effectively in terms of two elements: control or laisser-faire and rigidity or flexibility. Lastly, the third part of the answer concerns the project manager's competencies in these situations. Developing a plan based on rules and standards alone is no more effective than mechanically following a plan that may not necessarily lead to the project's success, especially in circumstances of great uncertainty. Does this suggest an emerging need for new competencies?

Chapter 14

Planning Risk and Cool Heads: Survival Conditions Required for Managing Projects

Jean Martel

When asked to be part of a symposium on polar expeditions and project management, my first reaction was not what most people would have expected. They would have asked: "What is the relationship between polar expeditions and project management?" I had received little detail regarding the exact nature of the symposium and was simply told to await further information by e-mail.

Suddenly and inexplicably, the realization in my little project manager's brain was immediate. I never could have imagined that two subjects could be so related at all levels. I had never associated these expeditions and project management, but they aligned perfectly. This symposium was a trigger for me, especially since I had also been invited to participate in a roundtable discussion following two days of presentations, where we would be asked to answer the question: "In your opinion, what are the essential qualities of a project manager?" We would discuss this topic and present our findings to the explorers and researchers. However, the topic itself is so vast that it is difficult to summarize, even if it was already clear in my mind.

I must also confess that in my particular case, I'm an explorer in my spare time. I am basically an athlete, having participated in several expeditions that I consider amateurish in comparison to those described by the expert presenters. But what are amateurs at their core? In the end, the nuance is rather small. Planning, danger, risks, all these elements are present in so-called amateur expeditions, just as they are present in "professional" expeditions. Following the analogy, these elements are equally present in small and large construction projects!

Perhaps it was simply by chance that I am a construction project manager by profession while also participating in amateur expeditions. These experiences provided me with the ability to understand both worlds: the explorer on an expedition and the project manager building a building, for example.

14.1. Expeditions and Projects

Before providing you with the basis of comparison between the two subjects, I present specific examples to justify my reasoning. First, I have participated in several different expeditions that took place in mountains, under water, on a mountain bike, and on a windsurfing board. During these expeditions, risks were ever-present. The weather conditions in Quebec can be extreme, on both water and land, and greatly increase risks. As the organizer of these expeditions, I had to plan, coordinate, and execute the expedition with my team members. I would, of course, delegate many tasks, but always retained the role of organizer and leader. Doesn't this remind you of the role of a project manager? In both roles I was responsible for either the success or the failure of the expedition (or project).

The construction projects I managed, or as I enjoy calling them, my "adventure projects," involved building amusement park rides. This type of construction was not common in Canada, and the type of equipment being installed was unique in the world. Doesn't this make you think of an adventure?

14.2. Analogy between Expeditions and Construction Project Management

My expedition partners often perceived me as the cornerstone of the team. Naturally, my qualities as a planner and organizer ensured that my team members felt confident in my abilities and tended to entrust me with the planning process, just as clients seek the trust of a project manager to complete their project. I realize now that my project management experience helped me to develop the personal qualities of an organizer, a planner, and a leader that are required to head an expedition, and vice versa.

An expedition is a project in itself and must follow the same five process groups of project management: initiating, planning, execution, monitoring and control, and finally closing (Project Management Institute, 2008). These five steps are essential for the smooth running of either a project or an expedition. From the initial idea of a project or expedition until their completion, several aspects remain the same.

14.2.1. Initiating

Initiating a project due to a market request or demand, organizing a scientific expedition, or planning even a simple personal challenge all share similar traits. Projects or expeditions must be initiated, have a beginning and an end, which makes them temporary and measurable [e.g., completion of a bridge with a budget (x) and schedule (y), or an expedition of 1500 miles to ski in Antarctica], and must account for the following factors.

- **Environmental factors.** Environmental factors can influence expedition or project risk. Not knowing water conditions (e.g., currents and water temperature) before sailboarding can have tragic consequences. Environmental factors can also influence the success of a project. A project in Africa is not carried out in the same way as a project in Canada. Cultural understanding is necessary for the proper functioning of a project.
- **Preliminary scope statement.** When defining a project or expedition, it is necessary to establish the bases of work, which is part of the first stage of the project—the planning stage. In project management, this step defines the mandate and is used to obtain financial backing. For an expedition, the scope statement allows the expedition members to secure sponsors. In fact, in both cases, this provides a preliminary definition of how things are to be done and what is to be achieved.

14.2.2. Planning

I do not believe that this topic needs to be further elaborated in this chapter. There is a familiar saying, "More planning results in fewer problems." In other words, planning is how to execute or achieve the project or expedition goals and is probably the most important step. The schedule and budget are also prepared in this process group. There is another analogy between project planning and expeditions. In the case of the latter, it is best to bring only what is strictly necessary. If the supplies are too heavy to carry or if equipment is missing, both of these situations pose an equivalent level of danger to an expedition. A similar situation exists in project management. There is no benefit to doing more than what is required, nor is there an advantage to doing less than is required, because the project will not meet its objectives. Planning establishes all the important fundamentals for the execution of a project or an expedition.

Risk Assessment

Before a project or an expedition begins, it is critical to identify risks. I regularly repeat to colleagues my strong conviction that risks must be eliminated. In fact, I accept a risk if, and only if, I have no other choice. During an expedition, this can be vital. Imagine being in an overturned kayak, in 4°C water, without either thermal protection or a rescue boat. What a disaster! This is a situation in which there is a great risk that one person panics and accidentally drowns his best friend who tries to save him.

In project management, risks that cannot be eliminated must be included in the budget as a contingency or reserve. If this step is ignored, you might see your best client try to drown you in order to save his own skin. This step is also important when dealing with a contractor. Have you ever seen the face of a manager of a construction firm, which has an idle-cost time of $5000 an hour, be forced to wait because of a problem caused by a poorly defined risk? His face is as twitchy and panicky as a kayaker in 4°C water.

A project manager is often the reason for success of a project but is also responsible for project failure. This role is equally important for an expedition leader. He can lead his team members and sponsors directly to success or failure (possibly even death). He must maintain his composure in all situations, never be overcome by panic, and never be like a kayaker in distress. Therefore, a project manager should never take on a task for which he is neither trained nor has the required experience; it is fundamental and even critical in the case of an expedition leader.

14.2.3. Execution

Following proper planning, establishing well-defined project or expedition goals, clarifying how works are to proceed, and evaluating the risks, a project manager or expedition leader passes to the next process group, project execution. It is at this point that money is spent the fastest and therefore it is important to follow plans. There is no longer any time to plan; it is time to run the project. One cannot stop a construction project at this stage, barring some catastrophic event, without impacting the budget and schedule. The same applies for an expedition: It requires planning. During its execution is not the time for the leader to ponder whether I should have brought a wet suit or foreseen the need for a boat. A solution must be found; it is critical.

Fundamental Principles

It is human nature to overlook something, no matter how much has been planned. This is true for both a project and an expedition. At any moment, even if you have the discipline to follow the plan throughout the project or expedition, you must possess the ability to adapt in order to save yourself.

A good example in project management would be working hand-in-hand with a contractor with whom you have signed a contract but who is unable to respect the project deadlines. If you "give it to him" without concern for the entire project, you risk that he may abandon the project. You will have the benefit of saying that you followed the principles of project management, and at the same time rant about the contractor's failings, but you will not be able to brag that the project was completed. Why not try to find a win–win solution? It is the duty of the project manager to ensure that all means have been considered to find a solution that allows the project to continue.

On an expedition, things can sometimes go wrong, even with the greatest of care used to identify risks and ways to mitigate them. Flexibility may be the difference between life and death. Why risk completing a 6000-meter climb if the weather forecast predicts bad weather for the next two days and a few members of your team are showing signs of extreme exhaustion? You have to be flexible, as an expedition leader, to the point of being willing to abandon the expedition until the following year without any hesitation to make the decision to do so.

14.2.4. Monitoring and Control

As a professional project manager, I can assure you that all of my clients want their projects to be successful, but they also want the budget (change control) and schedule to be respected. I was often asked, "Jean, are we on time and on budget?" by the U.S. owners of an amusement park where I had to execute unique projects never previously seen in Canada. As I had been hired to complete their project on the basis of my expertise in this field, they were more concerned about these aspects of the project than knowing whether the ride was going to work!

I believe that all expeditions should have accountability. Sponsors should get a return on their investment, without bad publicity. The explorers who discovered the Americas needed to be accountable. They needed to bring back proof of the riches of the new continents (e.g., gold and spices) in order to be allowed to return. The concept of a sponsor is not new. Even for smaller expeditions that you complete for yourself, you want a return on your investment.

14.2.5. Closing or Ending

This is a process group that is often forgotten or overlooked, but it is so important! It is during project closure that we complete the lessons learned that have been so carefully documented throughout the life of the project. The lessons learned will help us to better execute future projects. It is essentially the response to the following question: What would I do differently if I had to redo the project? When well documented, these lessons learned benefit both the client and the project manager when applied to other projects. This is also the time when we complete the administrative closure (i.e., contract and final acceptance of the completed product), as well as releasing resources retained for the project.

The similarities to an expedition are obvious. The lessons learned from climbing a mountain (e.g., information about roads, ice bridges, and slopes) can help future explorers to stay alive. They serve to reduce some unforeseeable risks. It is also at the end of the expedition that the scientific journals or articles are completed, and when we say goodbye to the team members.

14.3. The Essential Qualities of a Project Manager

Returning to the original question of this chapter, what are the essential qualities of a project manager?

Well, I would reply in the same manner as if I had been asked the question: What, in your opinion, are the essential qualities of an expedition leader?' In my opinion, the qualities of a project manager and that of an expedition leader are those that allow them to avoid a possible panic situation.

By keeping their composure and adopting a proactive, thorough, and flexible approach, project managers or expedition leaders can avoid high-risk situations. They will find themselves in a position to protect their team members, stakeholders, clients, and sponsors from a disaster. To do this, proper planning is required, accompanied by a comprehensive risk management plan.

Reference

Project Management Institute (2008). *A guide to the project management body of knowledge (PMBOK® Guide)* (4th ed.). Newtown Square, PA: Author.

Chapter 15

Flexibility and Rigidity in Planning a Program: The Case of the Montreal Metro Renovation Project

David Brazeau

I am responsible for budget and planning management of the Reno-System program whose budget is nearly a billion dollars, so I am confronted daily by rigidity or flexibility. The following presents, at a very high level, how the organization put in place for the Reno-System program adapts to rigidity and flexibility.

After a brief presentation of the Reno-System program, I discuss certain elements that cause rigidity in the program and those that offer flexibility. Finally, I discuss the desired balance put in place at the different levels of the organization.

15.1. The Reno-System Program

The Montreal transit authority, "Société de transport de Montreal" (STM), had been aware of the aging of the subway's fixed equipment, the vulnerability of its network, and the negative impact on the reliability and security of the service, and decided to set up the Reno-System program to renew the fixed equipment.

Begun in 2001, the program is made up of five-year phases, with a total budget for the first two phases of C$963 million. It involves the replacement or renewal of certain fixed equipment (e.g., escalators, wiring, and telecommunication system).

The purpose of the program is to ensure that the entire subway system operates reliably and safely. Its objectives are to maintain the reliability, maintainability, availability, and safety of the fixed equipment (thereby helping to maintain those of the subway network), optimize investment, and, finally, to improve customer service and the performance of the STM by benefiting from technological opportunities. Furthermore, interventions on fixed equipment are to be made without interrupting operations and minimizing inconvenience to the customer.

15.2. Rigidity

When planning the Reno-System program, many rigid aspects, including those presented in the following paragraphs, had to be considered.

The scope definition had to respect the *budgetary envelope* authorized by our financial partners. Rather than seeking to minimize costs to achieve a defined scope, we need to maximize the scope in order to meet the budget. The image that we often use to represent this feature is that of an aircraft carrier. The aircraft must land on the carrier, not before, not after. Thus, the budget allowed should be used to achieve maximum scope without generating cost overruns. This rigidity is counterbalanced by flexibility. In this case, flexibility is provided at the end of each execution phase, when the scope of work is adjusted to meet the amount of funds still available. This flexibility may have resulted in, for example, awarding certain smaller contracts at the end of the program. We can see that there is flexibility even when there is rigidity.

Another aspect considered was the *eligibility* to the grant. This issue concerned the type of equipment (fixed equipment only), the age of the equipment (end of life), the type of work (renewal or replacement versus regular maintenance), spare parts, and the acquisition of specialized equipment. This meant that opportunities to perform additional work in the same sector were possibly missed.

The *legal framework* took into account, for example, procurement (e.g., public tender, authorization levels for awarding contracts, approval by the Board of Directors of the STM), building and fire safety codes, the collective bargaining agreements governing the work done in-house, etc.

Of course, *operational constraints* had to be considered. Thus, a detailed plan was prepared to minimize the impact on customers, whether making escalators available in the up mode when work is performed on another escalator or by minimizing the work carried out in high-traffic areas and certain times of the year, such as festivals and the school year. The availability of sites was also considered: night work limited by sector, transportation of equipment in the tunnel, portion of the premises occupied by customers, among others. Routine maintenance and operational activities were also considered, as well as the delivery of other projects.

15.3. Flexibility

To implement a program of this magnitude, the project office developed and put into place an *organization*. Although this organization had to comply with certain rules of governance and align with existing systems, it was initially tailored to the needs identified. It was also able to evolve and adapt to the different stages of the program's execution. Its creation and evolution was made possible thanks to the flexibility that managers demonstrated in the management of the program.

Resource allocation is an element that provides flexibility in program planning. For example, project managers need not necessarily be skilled in the technical areas of their project, and it is therefore possible to assign a project manager who has the greatest availability. Another way we used was our partnership with a consulting engineering firm that allowed us to use *specialized resources* on an *ad hoc* basis.

During the execution of various projects within the program, a budget reserve was established so that amounts could be freed up during completion. This reserve allowed, toward the end of a phase, to select additional work that maximized the use of the entire available budget. A list of additional work was continually updated based on the priorities of the STM.

All project office *policies and procedures* were developed taking into account certain specific requirements, but they were mostly created based on the environment and the specific needs of Reno-System.

Aside from the organization that was put in place, managers also benefitted from flexibility in the *choice of locations and equipment* that were prioritized during the planning stage. This flexibility was particularly appreciated when opportunities or risks arise.

Indeed, different *opportunities* arose during project delivery because of other projects (e.g., the extension of a subway line) or other such addition funding sources as the federal government's transit security program. These opportunities can change the planning and even the scope of the program. The organization allowed required changes to be implemented within the framework of the policies and procedures developed for this purpose and in accordance with governance.

In addition to opportunities, *risks* can also lead to the revision of the plan and even of the scope. For example, some constraints related to the *location of equipment* led to a revision of the scope of the program (e.g., acquisition of land in downtown Montreal for the construction of ventilation stations), especially since several years are sometimes required to develop new scenarios. New elements related to *operational constraints* as well as the *socio-political environment* in which certain projects are executed also required a change in the program plan.

15.4. Finding and Maintaining Balance

The grouping of several related projects in a single program allowed the STM to achieve benefits that would not be possible if the projects were managed individually. The STM also ensured a higher level of control of the projects. These benefits were possible thanks to standardization of procedures, policies, and eligibility requirements.

Although flexibility was built into the development of the organization, this standardization could have represented rigidity for the people who needed to comply with it. These people had to use the flexibility available to them to fulfill their mandate. In some cases, they could put in place tools or ways of doing things that were adapted to their sphere of work and could create aspects of rigidity that they in turn imposed on others.

Flexibility can lead to the development of methods that can in turn be viewed as elements of rigidity. A balance between flexibility and rigidity is constantly sought at different levels: of program, projects, and services.

15.5. Conclusion

The planning of the Reno-System program was developed taking into account certain factors such as the rigidity of the budget authorized by financial partners, the eligibility of the work according to the grant, and the legal and operational constraints. Once these elements were identified, flexibility was available to develop an organization that was able to adapt to needs and the specific context of the program during execution. This flexibility also allowed planning and risk management to evolve and adapt during the various stages of program delivery.

The planning of the Reno-System program was constantly adjusted due to the flexibility available, although it was within a prescribed framework. This duality of rigidity and flexibility means that people at different levels of the organization must get involved in the planning to seek a balance to meet the needs of the program.

Thus, in the framework of the projects, a balance between rigidity and flexibility is always sought. Rigidity is almost always imposed. Methods must be put in place to compensate for this and to allow flexibility in the execution of the projects. Present in all levels of the organization, this balance differs from one person to another because flexibility that one creates can create rigidity for another. Although rigidity is generally imposed, flexibility is in turn is used to its full potential through the creativity of the different stakeholders.

Chapter 16

Project Manager: Specialist or Generalist?

Benoît Lalonde
Maude Brunet

As a practitioner and consultant in project management for GPBL, Inc., since 1996, I (B. L.) have observed that the range of practices in the field of project management is mostly similar from one industry to another. Indeed, GPBL helps many organizations in project management, whatever the industry. Project management is similar for a broad range of organizations, whether public or private, in industries as diverse as regulated, hydroelectric utilities, manufacturing, engineering, and corporate.

Though all organizations have a specific context and unique features, ranging from the sector of activity to its specific vision, mission, and organizational culture and structure, there is nevertheless an increasing and universal trend for organizational leaders to focus more on project management.

However, despite the growing interest in project management, companies often lack practical means and tools to improve their organizational maturity in project management. This chapter first outlines the main constraints faced by organizations concerning project management. Then, the role of the project manager is discussed according to two different perspectives: specialist or generalist. Finally, the main organizational enablers for improving organizational maturity in project management are presented.

16.1. Description

GPBL has developed an Integrated Project Management Approach© (Figure 16.1). This approach connects the strategic plan of an organization to project management in order to achieve the objectives desired and to fulfill the vision and mission. The cycle presented in this approach allows practices in project management to be improved, and the foundations of portfolio, program, and project management to be established. Following this, an assessment of the improvements and results enables an update of the strategic plan by top management, and the cycle may be repeated. In the midst of this cycle, a Project Management Office can play an important role in coordination. Also, some organizational enablers can foster optimal implementation of this approach. These factors are discussed in Section 16.1.3.

According to the Integrated Project Management Approach, the first step an organization undertakes to improve its practices in project management is to diagnose its current practices. To do so, GPBL favors the internationally renowned *Organizational Project Management Maturity Model* (OPM3)®, developed by the Project Management Institute (PMI). This model takes into account project portfolio management, program management, and project

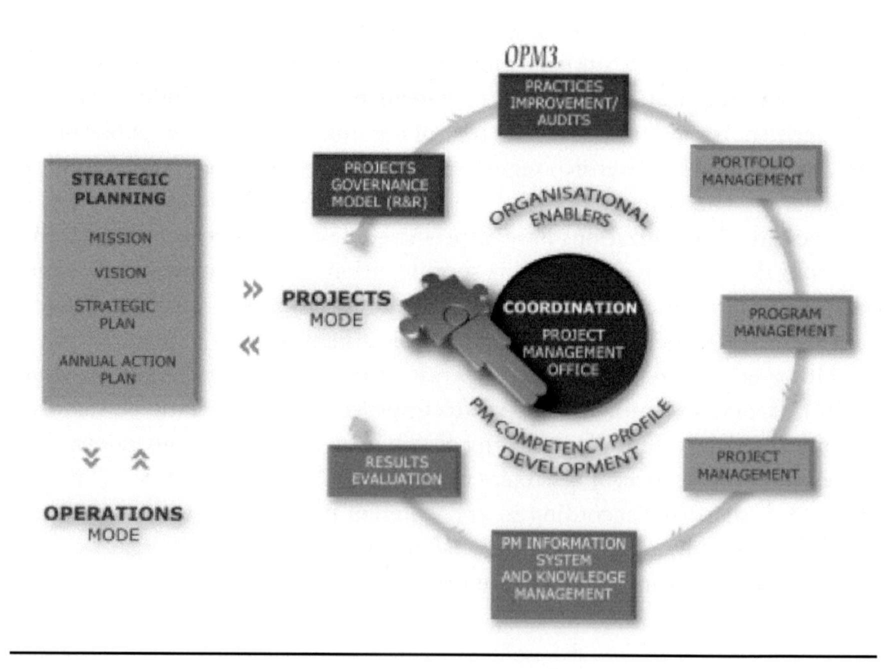

Figure 16.1 Integrated project management approach.

management. It has four levels of maturity for each of these management activities: standardize, measure, monitor, and continuously improve.

According to a 2008 survey by PriceWaterhouseCoopers, only 2.5% of companies at the international level carry out all of their projects. Additionally, more than 50% of projects end in failure. According to this survey, project failure is often caused by organizational factors such as the inadequacy of the processes, tools, and resources. These findings, as well as our own experience, show that few organizations, to date, are mature in project management.

16.1.1. The Main Constraints in Organizations

From our findings, organizations wanting to implement or improve project management do so in conditions that are far from optimal. Project managers are assigned to specific projects, often essential for the organization. However, many constraints make the work difficult, sometimes impossible.

We report in the following paragraphs the main constraints facing project managers. These constraints fall into three types: lack of top management support and commitment, lack of tools for project portfolio management, and lack of project management tools.

For project management to perform, it needs top management to be involved and understand the basics of project management. To select projects that can meet business needs, strategy must be thought out and then the vision and mission of the organization can be established. From this results the strategic plan and the annual action plan. Only top management can establish these plans to which all projects must be aligned. This exercise is too often overlooked by senior management of companies, or worse, it has been done but not communicated to all employees. Also, managers and leaders of organizations should understand the basics of project management. Project sponsors or clients for many organizational projects may lack knowledge in this field, which can lead to confusion in the definition of a project or misplaced and unrealistic expectations. In addition, a lack of training in project management generates misunderstanding about what project management is and what should be expected both from senior management and from all employees.

With regard to project portfolio management, few companies to date have the proper tools to manage and monitor their portfolios. This lack of tools has several adverse consequences, which affect prioritization of projects, which in turn leads to poor capacity management.

Although more and more organizations have adopted project management, many tools are still missing or inappropriate. Unrealistic constraints of time and cost can cause unnecessary pressure for the project manager, as well as the members of the project team.

16.1.2. Specialist or Generalist

The role of the project manager is still poorly understood in organizations, by both top management and employees. A project manager may be a manager or a professional. In many cases where a project manager is a professional, his authority is rather low and he is familiar with the contents of the project and is a specialist rather than a generalist. We question how effective this type of manager is in achieving project success.

Leaders need to understand the changing roles within organizations. A few years ago, specialists conducted the majority of their tasks in operational mode, in accordance with their job description. A minimal amount of time (about 20%) would be assigned to projects or mandates. Nowadays, these proportions have changed greatly. Specialists can now spend the majority of their time (70%) on projects or specific mandates, which leaves little time for operational tasks. Faced with these changes, leaders must recognize the importance of general project management and recognize the discipline as such.

Project management lacks recognition in the workplace. Too often, it is done intuitively by a project manager who knows the content but does not know how to manage the project itself, which leads to serious consequences. Leaders must recognize the role of project management and project managers. Under optimal conditions, project managers are generalists who are given a level of authority adequate for managing the project, and who work closely with content experts. Experts, or specialists, master the project product, but they are members of the team rather than the project manager. A project manager's main task is not to develop the content of the project, but rather to control costs, schedule, and scope of the project, while minimizing risks and maximizing resources. The current challenge is for business leaders to understand that a project manager must be a generalist, not a specialist. If senior management in organizations came to understand this fundamental difference, to recognize the required skills in project management and act accordingly, the results would be positive. Projects would be better defined, better managed, better controlled, better executed, and better closed. Our experience as consultants shows this. The establishment of a project management culture in an organization brings tangible and quantifiable benefits to the organization.

16.1.3. Organizational Enablers

When an organization's leaders decide to engage in project management, the first and most essential step is taken. However, the path leading to best practices can be long and arduous. Some organizational enablers can foster the establishment of a project management culture.

At the senior management level, the main actions to undertake are to align projects strategically, to develop a vision and an organizational policy for project management, and to create an appropriate organizational structure.

Human resources must develop a competency profile in project management, to conduct an assessment of individual performance, and if possible to develop a training program in project management.

A practical way to enhance project management in a company is to set up a project management office. Senior management must examine which are the best means to implement it to achieve its objectives. Nevertheless, several advantages can be identified by opting for the establishment of a project office.

To start with, a project management office can take a benchmark to determine what is done in the organization and beyond. Following these observations, action must be taken to strengthen project management in business by establishing organizational practices in project management, organizational methodology in project management, organizational techniques in project management, measurement tools in project management, and criteria for project success. In addition, other actions that may be taken at the organizational level include developing a system of sponsorship, mentoring, and coaching and developing a project management system. These measures allow optimal resource allocation and communication on projects. Also, setting up a project management community of practice and focusing on knowledge management can aid in maturing project management.

16.2. Conclusion

In conclusion, business leaders must be made to better understand what real project management and real portfolio management entail. Recognition of project management as a discipline helps project managers to improve their knowledge and skills. Organizations must move away from using specialists to using generalists to optimize project outcomes and benefits.

Recognizing the project manager's role as a generalist is important for the evolution of the profession toward best practices. Too many projects are still carried out by experts who master the contents but miss the essential constraints of time, cost, and quality. Organizations' top management should understand the critical role played by the project manager as an integrator and staff accordingly with people having skills in project management. Ultimately, projects and organizations depend on this for success.

Chapter 17

Project Management and the Unknown

Jean-Pierre Polonovski

My preferred domain is project and research and development (R&D) program management in the field of technology.

Of course, R&D projects cannot be entirely likened to expeditions, but they do have some fundamental similarities that are shared aspects of essential mechanisms implemented when managing both kinds of projects.

The characteristics inherent to such projects make them difficult to manage without using a collection of unconventional methods. In the following paragraphs, I discuss the fundamental principles of these projects and how they differ from traditional project management.

17.1. Shared Characteristics of Expedition Projects and R&D Projects

The main characteristics that expedition projects and R&D projects share are as follows:

- More unknown than known factors
- Identification of difficult risks

- High dependence on factors beyond one's control
- Very high dependence on individuals

Contrary to generally accepted ideas about project management, an R&D project cannot be entirely planned in advance. Furthermore, risks are often quite simply impossible to identify.

Many projects in more traditional fields do not receive prior authorization to proceed before all stages of project planning are explicitly defined. However, with R&D projects, it is essential not only to accept progress without knowing whether the project will succeed, but also to accept that this is inherent to the very nature of such endeavors. I am inclined to think this is equally true of expedition projects. What would an expedition be without associated risks?

To analyze the characteristics these two types of projects share, let us look at their similar outcomes:

- Both are entirely hit-or-miss.
- Their key feature is often creativity.

Specific circumstances arising in the course of a project must be taken into consideration. This is when creativity is important. These steps are often synonymous with the life or death of the project, if not of the participants. The distinctive feature of solutions in these projects is that they are often out of the ordinary, while also demonstrating discipline of thought resulting from real maturity. At the same time, these solutions are grounded in a global vision of the problem and its consequences.

17.2. Ambiguities of Project Management

The first ambiguity in project management lies in the notion of *ownership*. Often in R&D projects, the person in charge of the project is also the sponsor and creator of the project. In more traditional projects, the project manager is "parachuted" into a pre-established project.

By definition, a project has an end. In a certain way, a project manager therefore regularly finds himself (or herself) in a situation of job loss. It would be difficult not to find this situation somewhat insecure.

The notion of loyalty is also central to the discussion. Matrix-based organizations emphasize the duality of project participant loyalty. Will they favor the department they work in generally, or the project in which they participate? In French, the term *gestionnaire* (organizer or manager) is employed; in English, "manager." But what does a project manager really *manage*? People, a budget, suppliers. . . ? Compared to an operations manager, for which profile the employee's professional history is key, a project manager's success is measured by the manager's ability to undertake a project, his behavior, and the conformity of the methods he employs. Here again, the notion of a global vision is important.

A project manager must also distinguish between clients and suppliers. Will she, when it is not in the immediate interests of her employer to do so, take the side of a supplier and consequently bite the hand that feeds her? Does this not constitute a conflict of interest?

In other words, the very task of a project manager is ambiguous and complex. It requires much judgment and maturity to deal with such complexity; a wealth of resources and discernment must be implemented.

17.3. An Indispensable Factor in the Success of a Project

Let us first define what the success of a project is. In R&D projects, but also in expeditions, I believe, success is not always about achieving the original objective, but to some extent, surpassing certain limits. It is about going beyond what is known and into the unknown.

I do not know of any project having been successful in these domains for which the roles and responsibilities of each member were not particularly well understood or even if necessarily made explicit. We have heard of situations in which roles and responsibilities are attributed according to a hierarchy or to a particular field of expertise. The latter situation is especially frequent in R&D projects.

In this book there are case studies evoking an analysis of battle scenarios. Completing a project in this context, particularly when unforeseen events arise, is often achieved because each participant knows his or her role, and prior confidence in each other means nobody questions the capacity of the decision maker to make the right decisions.

We can, however, distinguish two types of roles and responsibilities: individual ones, which must be clearly defined and whose consequences must be clearly understood, including their implications; and collective ones.

The notion of accountability is also key for individual participants. It goes hand in hand with the recognition that is gained in the event of success. Accountability is also a reassuring factor within a team context.

Concerning collective roles and responsibilities, a major factor of success can be described by the famous phrase: "One for all and all for one." However, in this case, the "all" in "one for all" consists of describing the contribution of each individual to the collective success of the project. The "one" in "all for one" represents the objective to be achieved.

In such fields as R&D, and certainly in expeditions, easy-going personalities cannot be counted on. On the contrary, very strong, sometimes unsociable characters are often encountered. The collective reaction to this in such contexts often occurs through explicit recognition of each person's expertise. Faced with unknown situations, this recognition enables attention to be naturally directed to the expert in the relevant field. Consequently, decision-making roles and roles of expertise are often dissociated. What is expected of a project manager here is an ability to listen, lead the debate, summarize discussions, and finally reach a decision.

A natural distribution of roles and responsibilities, when time is available to do so, will often reflect a combination of such expertise with the quality of communication skills of each individual within the group.

The more critical the unknown situation is, especially concerning reaction time to save the project's (or one's own) life, the more prior acceptance of roles and responsibilities is crucial. Herein lies a potential source of conflict. In a critical situation, conflict results in failure. R&D and exploration projects are quite different from the military missions discussed in this book. Indeed, discipline, the very core of military life, is not applicable here. Consequently, roles cannot be attributed according to codes, but must be negotiated and gained through merit, and accepted by fellow participants.

17.4. Flexibility and Control

The fundamental task is to describe the fundamental qualities of an expedition project manager.

As stated previously, it is important to remember that the nature of project management depends on the domain in which it is implemented. While overarching aspects remain constant, strategies to meet specific aspects of the domain in question must be implemented.

With this in mind, are we looking for a project manager who is flexible or controlling?

Before answering this question, I would like to describe characteristics of highly competent project managers I have encountered:

- Adaptability and ability to anticipate
- Capacity to estimate the consequences
- A perfect understanding of the global picture
- Strong leadership coupled with an equal capacity to re-evaluate oneself and the situation

As discussed previously, the task of project managers is ambiguous and complex, requiring much judgment and maturity. An undeniable differentiating factor is the ability to maintain a global picture. This enables actors to remain focused on the essentials, especially in times of crisis.

Leadership is necessary because authority must be earned; it is not dictated by rules. However, this leadership must be accompanied by a strong ability to listen and to give recognition and praise. A critical situation is often characterized by the fact that expected solutions have not worked. In such circumstances, leadership must involve evaluating strategies and openness to innovation possibly coming from unlikely sources.

So, flexibility or control? Project sponsors and participants alike must find a working human balance of the two.

17.5. Management and Creativity

It seems rather a contradiction in terms to want to manage creativity. Nevertheless, creativity must be channeled in order to be aligned with the objectives and stakes of the project.

In the case of R&D projects, as with expedition projects, creativity occurs before the project is undertaken; it is the very foundation of the project. Indeed, for each project, new hypotheses must be put into practice. However, the point I would like to make here concerns the creativity required to resolve crisis situations.

There are two types of crisis situations. There are military interventions, in which reaction time is minimal and consequently reflection is out of the question. In these situations, it is anticipation of different scenarios that opens the door to creativity, as much in generating the list of possible scenarios and appropriate reactions. This is a sort of training to overcome any inappropriate instinctive reactions. This situation can be resolved by the idea that all has been done to avoid encountering unknown situations.

I do not doubt that such situations are met in the course of expedition projects, and cases offering the opportunity to reflect and evaluate the situation are also frequently observed. It is indeed necessary to control panic, to be able to step back and evaluate a situation when instinct tries to take over. A large part of such reflection consists of eliminating unknown factors.

This observation nevertheless resides in a context in which unknown factors are expected. Another approach is possible. It is indeed not a common approach in the world of project management, but it does provide undeniable advantages for R&D projects. Unknown factors can thus first be accepted, then evaluated, and finally used to the advantage of the project where possible.* For this, it is crucial to achieve a significant psychological phase: accepting mistakes. Error does not equate to failure. An error does not necessarily risk the whole project but may damage the prospect or outcome of optimal solutions.

In such situations, one must often react by improvising. For such improvisation to be accepted, it is essential that participants have a shared *vision* of the project. Remember here that communication is the crux. This must not be reduced to merely communicating results or specifications or even management of interpersonal exchanges. The fruit of communication must combine transparency, identification, and a sense of belonging to the group, and a shared vision among members.

Two significant factors of stability in crisis situations reside in taking responsibility and employing democracy. Decisions should not be taken alone, in an authoritative manner; a project manager must rather direct decision making, accepting to decide between solutions proposed.

The human factor is important: Recognizing and accepting differences is essential.

17.6. Conclusion

The natural tendency is to want to compare and contrast opposites. The themes proposed in the conference are as follows:

- Flexibility–Control
- Specialized–Nonspecialized
- Planning–Composure
- Flexibility–Rigidity
- Management–Creativity

Jean Martel said that project management is like life. The metaphor of the rearing of one's children immediately comes to mind. In life, do we want parents who are flexible or controlling? Life teaches us that flexibility is necessary to be able to flourish, while control is required for safety and learning.

As discussed in the first sections, the task is ambiguous and complex. It must be undertaken by using a wealth of both resources and discernment. This wealth consists of the wide range of resources made available to project managers. Discernment comes from within the project manager.

* Thierry Picq, *Miser sur l'imprévu*. In the present context, the term "unknown" is more appropriate than "unforeseen." Here we are talking about circumstances that could not have been foreseen.

Coming back to the metaphor of parenting, we want parents who are both fair and critical. Above all, we want parents who encourage us to exceed ourselves. The essence of expeditions similarly lies therein.

This chapter also discussed the difficulties presented when trying to find solutions in critical situations, and the importance of a global vision. I do not want to end this piece without discussing the key mechanism, which is transmission of experience.

I am referring to experience, and not knowledge (for which transmission techniques exist). Knowledge is the essence of the whole schooling system. Experience is sometimes synonymous with the irrational: Experience is gained only through openness to others, to their differences and their uniqueness, and by the capacity we must develop to receive and accept.

Acknowledgments

I would like to thank all those who have collaborated on these expeditions. At a time when we have everything, when technological means enable us to overcome nature in many domains, expeditions might seem futile. These people nevertheless show courage, because there is no courage without consciousness of risk. So I take this opportunity to praise all those who have shown how to push oneself to the limits.

Chapter 18

Control and Flexibility: Which Balance Do We Mean?

Danielle Desbiens

What does "balance between control and flexibility" mean for project management? As a long-time psychologist, tenured professor at a business school, and project management team coach, I base my answer on an exploration of the different types of environments in which a project leader works: the project environment, the team environment, and the individual environment. Given the complexity of team environments, I believe that project leaders must act according to various perspectives depending on whether they manage a team, manage with a team, or manage as a team. In doing so, balancing control and flexibility requires specific diagnostic and learning competencies, which is the second theme of this chapter.

18.1. Which Balance Do We Mean?

Let's start by defining terms. Control and flexibility: Are they truly opposites? Are they opposite ends of the same continuum? Control is often seen as a factor that paralyzes or blocks initiative. The negative connotations attributed to the word "control" are based on the assumption that flexibility is its opposite. Accordingly, balance is perceived as the midpoint between two extremes. However, control and flexibility both vary by degree, and increasing one does not mean eliminating the other. They do not belong to the same continuum, but rather comprise two separate functions. The opposite of control is laissez-faire, and the opposite of flexibility, rigidity. (See Figure 18.1.)

In project management, both control and flexibility are necessary. A project leader must use them like both hands. Like the captain of a ship, a project leader checks the current position against the chosen course often, steering the ship according to the dictates of wind and current. Without such control, the precise position is a matter of conjecture. The metaphor of the captain highlights that control, in terms of monitoring, is a positive force. It provides a project leader with information on the condition of a system or part of a system, to check whether it meets certain desired criteria. Flexibility is also needed to adapt to various environments, not to mention its role in the application of controls. Based on expertise and experience, with or without consulting crew members, a captain reads each situation and makes decisions. The captain's flexibility depends on the leeway inherent in the situation at hand and its many variables, such as the degree of urgency, issues, and time.

A project leader's balancing act between control and flexibility calls to mind the concept of "ambidexterity" (Aubry & Lièvre, 2011). To take the notion of

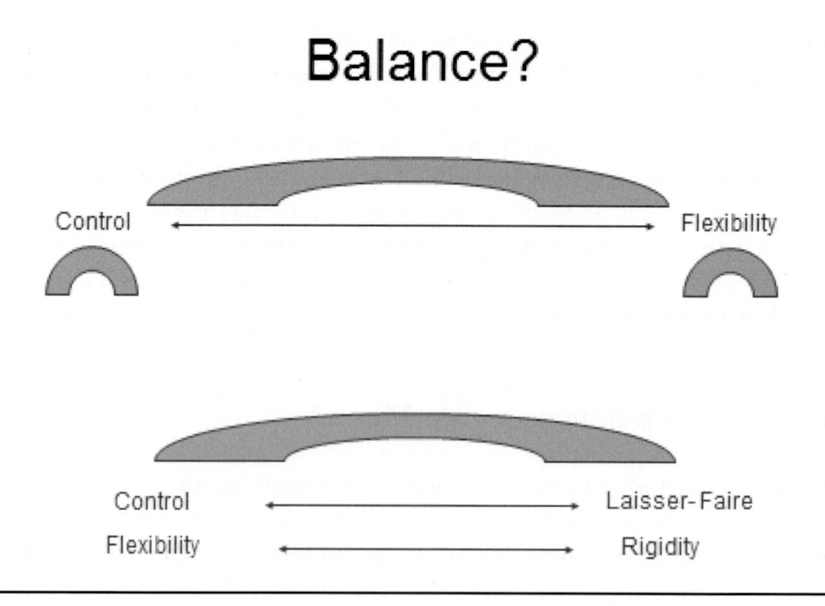

Figure 18.1 Balance or continuums between control and flexibility?

balance one step farther, this chapter poses the question: When would a project manager use control and flexibility? The immediate answer to this question is, "It depends." To clarify the matter, this chapter considers a few problematic areas by exploring the contingencies of management in three types of environments: the project, the group, and the individual.

18.2. The Project Environment

A project comprises a series of actions undertaken to achieve a specific outcome. It always involves some elements of change and innovation. The notion of a point of no return (Midler, 1995; Lenfle & Midler, 2003), in the literature on innovation offers interesting insight into the project environment. Based on the case study of Renault's development of a new car model, Midler showed how the margin for maneuvering diminishes as knowledge of the product increases. Once knowledge is sufficient, a project is "locked in," or in other words, the point of no return is reached. Some decisions are no longer to be questioned. According to Midler, therefore, flexibility is limited.

This notion of lost flexibility, associated with a narrower margin for maneuvering, aptly applies to technical projects. But does the same phenomenon apply to an expedition or any type of project? The moment a ship leaves the shore or a team heads off to the mountains can be considered a point of no return, as in the project studied by Midler (1995, 2003). In other words, sufficient knowledge exists to take the risk of leaving. However, at the same time, control over several variables begins to erode in relation to the environment, and there is a constant ebb and flow of the expected and unexpected. Control mechanisms remain active. To remain "in control," an expedition leader and members of the team read their environment while gauging the risks. Managing the variations caused by temperature, storms, equipment failure, or other contingencies, which Hällgren (2007) calls "deviation management," also demands considerable openness, flexibility, and creativity. Therefore, the control and flexibility functions are both present at every stage, before and after the point of no return. The possible existence of a new margin of maneuverability, rather than a reduced margin of maneuverability, warrants further investigation.

18.3. The Team Environnent

The complexity and scope of projects make them a team affair. A project leader has a different perspective. This implies knowing when and how to *manage a team*, *manage as a team*, and *manage with the team*. These three levels of management can be described in different ways. The nuances are subtle. Differentiating among them provides points of reference for a better understanding of change in team-related activities.

18.3.1. Managing a Team

A group's evolution is part of a growth process leading to cohesion. A project leader must establish mechanisms that encourage a team's development as an entity. A leader works to create a system that learns and performs effectively, can function amid uncertainty, and adapts to the movements of the others like the members of a sports team. This consolidation work occurs amid the tide of changes at work in a group. A team is not a stable environment. Like a person, it is born, grows, and dies. It can learn and it can regress. It can be reliable, strong, and relatively cohesive. A team is sensitive to any form of modification that does not originate within the team itself. It has its own dynamics. A project leader can influence these dynamics but not necessarily control them. A leader can:

- Understand and anticipate phenomena
- Determine the structure, specify roles
- Coordinate the efforts of members
- Affect group processes by applying controls

However, a leader cannot rigidly maintain the methods of operation put in place, and must deal with the team's dynamics by remaining flexible.

18.3.2. Managing as a Team

Managing as a team is possible because each element of the well-known PODC (plan, organize, direct, and control) process can be performed in whole or in part as a team. Depending on the needs and expertise of each individual, a project leader can work on these management activities together with one or several team members. In this way, a leader gains a better understanding of the situation and can enlist collective creativity to find solutions that reflect diverse points of view. The final product is not necessarily better, but the approach improves the probability of greater support.

A group is not a complete and perfect entity. Managing as a team entails its share of constraints. This management method is time-consuming and requires coordination. A project leader must deal with the group's strengths and limitations alike. For example, managing as a team involves the risk of encountering such phenomena as the escalation of commitment and groupthink. In the first instance, a group clings to a decision even though a dead-end outcome is a foregone conclusion. In the second instance, a group reaches a consensus or pseudo-consensus too effortlessly in the aim of protecting cohesiveness at any cost. Management as a team is not limited to critical periods alone. It comprises a combination of consultation and decisions both individually and as a group. This may be a factor of a project leader's preferred management style.

18.3.3. Managing with a Team

Managing with a team means making optimal use of the resources team members offer. It involves exercising shared leadership. Like an expedition, project expertise and experience originate with members of a team and, as experts, each member can influence the task at hand. Each member also influences how the group functions. Each member can formally and informally encourage the development of team spirit. In keeping with the spirit of the "high-performance team" model developed by Katzenback and Smith (1994), members work to attain the objectives of each of their colleagues to the same degree as they work toward the project's overall success. To achieve this, they must keep sight of and be able to communicate their own needs and the needs of others. They act in a spirit of collaboration rather than competition. Although it is demanding, and although optimal functioning is not always possible, managing with a team is an unavoidable project management necessity.

18.4. An Individual's Environment

An individual's environment is based on the uniqueness of each member individually, including his or her desires, fears, and moods. Regardless of a team's cohesiveness or the strength of its collective "we," a project leader cannot ignore individuals. They are the source of a group's energy. Among all other concerns, a project leader must also give constant priority to assessing the level of team members' motivation. A project leader takes the team's "pulse" and is responsible for helping everyone stay on course and maintain his or her level of energy. Nothing can be taken for granted at the individual level. Everything is a matter for attention and concern. Emotions are changeable. Moments of discomfort and high tension often unleash unpredictable reactions. To confront this challenge, there are as many approaches as there are individuals, and being fair does not mean being "always the same." One approach may work with one person and fail with another. Empathy and listening are instruments of choice.

Leadership is situation-dependent, adapting to the constant changes in individual issues. It is also transformational and concerned with sustaining energy.

18.4.1. Competency

This chapter's analysis of project context, teams, and individuals shows that a project leader must recognize the steps involved in a project, particularly the point of no return, show competency in strengthening the team, be able to share leadership, and know how to pay attention to individual problems. Lastly, a project leader must act in all of these areas without losing legitimacy as the "leader." Regardless of the environment, a project leader must adapt controls and exercise flexibility and must be "ambidextrous," which involves developing two competencies: a diagnostic competency and a learning know-how competency. A diagnostic competency is needed for a project leader to constantly see and recognize phenomena, assess issues, and decide what action is to be taken. If a leader lacks this expertise, she or he must at least be able to understand and gauge the merits of the response of the team's expert and know when to do the same for the entire team's response. The way that a project leader uses the expertise of group members depends on such other underlying variables as self-confidence and a sense of professional efficacy. The competency model developed by Spencer and Spencer (1993) views competence as the outcome of the dynamics among five elements: motives, traits, self-concept, knowledge, and abilities.

Competency can be defined as a combination of knowledge, know-how and social skills, in the exercise of a specific employment. Because of the uncertainty with which each project is carried out, a fourth type of knowledge is needed to become and remain competent: "know-how to learn" (Desbiens, 2005): "Know-how to learn" requires awareness of one's learning style and of the obstacles created by certain habits and attitudes. In project management, the practice of reflecting on lessons learned is well known but usually occurs at the end of the project. From a "learning know-how" perspective, this should happen on a regular basis with respect to practices, successes, and failures, and, per Argyris and Schön (1996), in single and double loops.

The boundaries between various project management environments are porous. Points of contact exist for each boundary. Wherever a leader intervenes, regardless of the level, the leader's essential legitimacy is at issue. Identified by Garel and Lièvre (2011) in an example on decision making following a difficult crossing, this concept is rich in meaning and importance. It encompasses the confidence that team members must have in their project leader. According to the literature, more than ten behaviors elicit trust. It can be summarized by the concept of dual credibility (Desbiens, 2010): professional credibility and relationship-based credibility. Professional credibility is earned on the basis of recognized expertise and experience; relationship-based credibility refers to behaviors of honesty, consistency, and authenticity (i.e., "walk the talk").

18.5. Conclusion

This chapter concludes with the following remarks:

1. In project management, control is essential, and may even be considered the "value added" of a project manager's work. Flexibility is useful in light of the uncertainties and shifting circumstances in which a project leader works. The concept of ambidexterity aptly summarizes the need to enlist both of these functions.
2. In this chapter, control was approached primarily as an informative aspect of monitoring and flexibility as an adjustment to the knowledge gained. In future research, it would be interesting to develop continua adapted to project management. Would they remain the same if control were associated with the power relationship and if the rigidity component of flexibility were considered the enforcement of rules and principles?
3. The requirements of work in various environments begs the conclusion that ambidexterity, defined as the concomitant use of control and flexibility, be considered a competency. For a better understanding, would it be helpful to examine the role of diagnostic competency and of the "know-how to learn" competency? How do a project manager's personal traits enter into the equation?
4. Management of a team, as a team, and with the team consists of transforming individual energy into group energy, fostering synergy, and reducing the negative impact of certain group phenomena. The key elements of this complex undertaking can be summed up in the following sentence: *Being part of a project that we own provides inspiration*. Everyone involved in project management should grasp the deeper meaning of this sentence.
5. This topic is extremely promising and its applications are numerous. In addition to management applications, such as the ones identified by Hallgren (2007) in analyzing deviations, it could also provoke thought among event management specialists. Many teams participate in projects that share features similar to those of expeditions and operate under the whims of circumstance. Is the point of no return the same? At what stage does the adventure begin?

References

Argyris, C., & Schön, D. (1996). *Organizational Learning II.* Boston: Addison-Wesley.

Aubry, M., & Lièvre, P. (2011). L'ambidextrie comme compétence des chefs de projet. In *Gestion de projet et expéditions polaires, que pouvons-nous apprendre?* (chap. 2). Québec City: University of Québec Press (PUQ).

Desbiens, D. (2005). Apprendre à apprendre avec la méthode des cas. In M. Bédard, P. Dell'Aniello, & D. Desbiens (Eds.), *La méthode des cas, guide orienté vers le développement des compétences.* Montréal: Gaëtan Morin Éditeur.

Desbiens, D. (2010). Contrôle et flexibilité. De quel équilibre s'agit-il? In Aubry, M., & P. Lièvre (Eds.), *Gestion de projet et expéditions polaires, Que pouvons-nous apprendre?* Québec City: University of Québec Press (PUQ).

Garel, G., & Lièvre, P. (2011). Le projet d'expédition polaire et la gestion de projet. In *Gestion de projet et expéditions polaires, que pouvons-nous apprendre?* (chap. 1). Québec City: University of Québec Press (PUQ).

Hallgren, M. (2007). Beyond the point of no return: On the management of deviations. *International Journal of Project Management, 25*(8), 773–780.

Katzenback, J., & Smith, D. (1994). *Les équipes haute performance.* Paris: Dunod.

Lenfle, S., & Midler, C. (2003). Management de projet et innovation. In H. Mustar and P. Penan (Eds.), *Encyclopédie de l'innovation.* Paris: Economica.

Midler, C. (1995). "Projectification" of the firm: The Renault case. *Scandinavian Journal of Management, 11*(4), 363–375.

Spencer, L. M., & Spencer, S. M. (1993). *Competence at work: Models for superior performance.* New York: John Wiley.

Conclusion

Gilles Garel

We are now able to draw specific conclusions about polar expeditions in particular, as well as some general conclusions in terms of project management at large. The conference first helped to define polar expeditions as projects—to transform this "topic" from something viewed as exotic into an object of serious research for management science and a learning source for managing "extreme situations." Initially, this was a paradox, because polar literature was never considered to be project management.

If we did not take into consideration polar logistics research (Lièvre, 2003), we could conclude that the topic of polar expeditions was absent from project management literature (Kloppenborg & Opfer, 2002). At most we could find some metaphorical allusions or pedagogical applications (Koehn, Helms, & Mead, 2003), but a polar expedition was not a serious subject in project management research. Two major explanations may be advanced. One, researchers do not typically have access to polar expedition situations. Two, polar expeditions would be considered as outside the project management field. How can we state that a polar expedition is a project? There is ambiguity about the output of a polar expedition project. In comparison, new product development or services are certainly more traditional situations for project management researchers. What about a sea kayaking expedition in Greenland, for example? What is the deliverable of such a project? The project output is intangible. Could the memories of the expedition, the "good times" spent together and shared by the team, be considered seriously as output? In other words, is it a "real" project? Is it serious matter, or just a vacation?

An expedition involves travel, mission, and equipment. A historical perspective shows how the "lineage" of expeditions led to the exploration of new territories. The project is the organizational management of the expedition. In practical terms, "project" and "expedition" have merged, and we talk about "expedition projects." In classical types of project management, polar expedition projects would be considered as "event projects." According to Arctic specialists, polar expedition projects can be categorized as "sporting achievement," "skiing fun," "exploration discovery," and "scientific research" (Lièvre, Récopé, & Rix, 2003).

Exploration is the gradual discovery of a new world. The degree of novelty is defined by the actor's point of view (i.e., "I can explore something new for me but already known by the rest of the world."). Such early polar expeditions as the conquest of the poles or the search for a Northwest Passage were full explorations because the participants were the first to enter these territories. Ships explored new and previously unknown areas (new not only for their crews but also for the world). During these explorations, the crew had to invent new devices adapted to the singular situations they faced and had to manage. In others words, to explore successfully, they were driven to innovate in project management.

These chapters have shown that all the characteristics of a polar expedition define a project, from upstream and preparatory phases to the actual project implementation phases, and finally to the postproject capitalization phases. Furthermore, this activity is temporary, specific, combinatorial, and uncertain. A polar expedition project is a temporary organization (Ekstedt, Lundin, Söderholm, & Wirdenius, 1999; Lundin & Söderholm, 1995) managed under the constraints of time, cost, and space, as part of a specific task, and it combines issues and personal agendas of the actors. There is a need for coherent organization, oriented toward a single goal, and requiring tools to control and plan. There is the research of the Scandinavian School into temporary action-oriented organizations, which are mobilized around an analytical framework for defining a mountaineering project (Hällgren, 2007). According to this school, such a project cannot be reduced to the implementation of tools, but it is an organization in itself. Such a project is sometimes created for ambiguous reasons that may not be rational, and may include personal ambitions or desires. So, polar expedition projects are full projects! They share characteristics with more generic forms of projects (Midler, 1996). The output of a project in the case of "sporting achievement," "pleasure of skiing," and "exploration discovery" is unique with respect to projects traditionally studied in the literature. This output is a service that is consumed as the project unfolds, as well as ex-post as a set of deep memories and learning experiences to reproduce in a similar project or to be used in different projects. For the team members of such projects there is also the pushing of personal boundaries at the intersection of discovering nature, autonomy, and physical challenge, whether or not they were striving for performance. This echoes the characteristics of patterns of engagement in a project team (Picq, 2005). Many studies focus on the meaning of the project for team members, the need that a project must make sense for those who are engaged in it. With polar expedition projects, this question of meaning is determined by the nature of the project itself!

Defining an expedition project as a project in itself makes the "expedition" an item to manage. By defining an expedition project as a project in itself it is also an object of the expedition, thus making it relevant for project management research. Apart from this foundation, this book taught us about the risky and ephemeral nature of these projects, their leadership, and the necessary adaptive learning before, during, and after these projects. This book is an act of recognition of the foundation of polar project management. This exercise has helped to clarify related vocabulary and legitimize this topic in the field of project management.

References

Ekstedt, E., Lundin, R. A., Söderholm, A., & Wirdenius, H. (1999). *Neo-industrial organising, renewal by action and knowledge formation in a project-intensive economy.* London: Routledge.

Hällgren, M. (2007). Beyond the point of no return: On the management of deviations. *International Journal of Project Management, 25*(8), 773–780.

Kloppenborg, T., & Opfer, W. (2002). The current state of project management research: Trends, interpretation and predictions. *Project Management Journal, 33*(2), 5–18.

Koehn, N., Helms, E., & Mead, P. (2003, April). Leadership in crisis: Ernest Shackleton and the epic voyage of the *Endurance. Harvard Business School Cases* (pp. 1–41).

Lièvre, P. (2003). *La logistique des expéditions polaires à ski [Logistics of polar ski expeditions].* Paris: GNGL Productions.

Lièvre, P., Récopé, M., & Rix, G. (2003). Finalités des expéditeurs polaires et principes d'organisation [Aims of polar expedition members and organizing principles]. In P. Lièvre (Ed.), *La logistique des expéditions polaires à ski* [*Logistics of polar ski expeditions*] (pp. 85–101). Paris: GNGL Productions.

Lundin, R. A., & Söderholm, A. (1995). A theory of the temporary organization. *Scandinavian Journal of Management, 11*(4), 437–455.

Midler, C. (1993). *L'auto qui n'existait pas; Management des projets et transformation de l'entreprise [The car that did not exist; Project management and business transformation].* Paris: InterEditions.

Midler, C. (1996). Modèles gestionnaires et régulations économiques de la conception [Management models and economic regulation of design]. In G. De Terssac & E. Friedberg (Eds.), *Coopération et conception* [*Cooperation and design*] (pp. 63–85). Toulouse, France: Octares.

Picq, T. (2005). *Manager une équipe projet* [*Managing a project team*]. Paris: Dunod.

Epilog

Pascal Lièvre

As Gilles Garel noted in the Conclusion, one of the contributions of this book is to build up the scientific legitimacy of research into polar expeditions in management science. This legitimacy was furthered when Brian Hobbs, head of the Project Management Chair at the School of Management Science at the University of Quebec at Montreal (ESG-UQAM), agreed to organize a symposium (on which this book is based) dedicated to research in this area, which gave recognition to its scientific value. This event demonstrated that expeditions on land, on the sea, and in the, air can be considered as projects, in line with the definition by Midler (1996), and that it is possible to see them as "event-projects in natural extreme environments." I propose now to place in perspective this symposium, in the broader context of the research program that I have run for more than ten years, "Managements of Extreme Situations," whose field of reference is polar expeditions (Lièvre, 2005).

Polar Exploration as a Field Benchmark

In December 2000, I participated in the organization of a symposium at the Maison des Sciences de l'Homme in Clermont-Ferrand, France, in collaboration with the University of Aix-Marseille II. The theme was "Logistics in Extreme Environments" (Lièvre, 2001), which sought to identify organizational principles in the field of polar expeditions. Two important ideas inspired this symposium. First, a new class of management situations is gradually emerging in organizations, characterized by high complexity and hyper complexity that Edgar Morin characterizes by uncertainty, risk, ambiguity, volatility, and the emergence of the new, which apparently does not fit the classic corpus of management science. Second, some actors have been forced for centuries to confront the complexity of certain situations such as polar expeditions. Explorers have developed, pragmatically and through trial and error, suitable organizational practices from which can be taken lessons about management. The field of polar expeditions has become a very rich area for learning how to conduct projects in an evolving, uncertain, and risky environment—what is called "extreme situations management"(Garel & Lièvre, 2010; Lièvre & Rix-Lièvre, 2013).

We can now take stock of achievements in this area of research. Advances include the following:

- The legitimacy of considering polar exploration as projects
- The question of the nature of actors' commitment and the motivation involved in projects
- The role that social networks play in mobilizing experts in the start of projects
- The need for organizational ambidexterity in project teams throughout the project life cycle
- Construction of a device, relying on organizational ethnography, to investigate projects

Because of our work and published research, we do not believe that the issues we discuss in the polar expeditions field are far from those in management science. We could say the contrary. In addition, this research that has been conducted for longer than a decade and has opened up new prospects for development. All this shows how we benefit from work in the same field over a long period. Our work also developed in other areas related to polar expeditions, such as fire and rescue services (Gautier, Lièvre, & Rix, 2008; Lièvre & Gautier, 2009), expeditions in high mountains (Lièvre, 2012), and space exploration (Bonnet & Lièvre, 2014).

Approach-Based Practices

We recognize the field of practice as autonomous and irreducible to any theoretical reduction area (Schön, 1983). This is within the framework of an epistemology of collective action (Hatchuel, 2000), where we take the usual practices of the actors as a legitimate object. This means investing what actors do and how they do it. This approach is focused on practice and has produced research in English and French under the name "strategy as practice" (Jarzabkowski 2005; Rouleau, Allard-Poesi, & Warnier, 2007; Rouleau, Chanal, Golsorkhi, & Langley, 2007; Whittington, 1996). There are also projects that are the subject of "project as practice" (Blomquist, Hällgren, Nilsson, & Söderholm, 2010). This perspective is closer to what some call "the practice turn" (Schatzki, Knorr Cetina, & von Savigny, 2001), which originates in the work of sociologists such as Bourdieu, Giddens, and Latour, as well as psychologists such as Piaget or Vigotsky.

An Appropriate Methodology for Understanding Practices

An issue in this area of research is developing methodological tools of investigation that conform to what we have defined as a practice. Indeed, documenting the "simple" progress of a project by focusing on the practices of actors rather than on their talk about their practices is not a trivial matter. Over several years, Rix-Lièvre and I have built a device for this type of investigation (see Chapter 5). This is a video device used for organizational ethnography that was developed through identifying a number of methodological problems. One of these problems is raised by Midler (1996) and is related to the presence of a researcher in an organization. A researcher should be accepted in principle by an organization, which cannot be taken for granted. Two related obstacles emerge, and they are related to confidential strategic elements for an institution in question (this is more consequential for an innovative company in a highly competitive market) and the confidentiality of certain information. The second problem is about understanding "doing" in detail. There are typically two ways to understand this: We can try to account for "doing" either through "saying" or through "observing." Both modes used independently raise questions. Thus, many studies have shown the inherent limitations of simply investigating "doing" through "saying" in the field of sociology (Peretz, 2004) and in management sciences (Argyris, 1993). When mobilizing ex-post through conventional interview techniques from sociology, it appears that an actor attempts to make his or her action legitimate vis-à-vis what is said. As noted on numerous occasions by Argyris and Schön, there is a world between action theories professed by actors and their theories in practice. Research from the psychology of work shows that it is difficult for an actor to simply describe the actual course of an action, because knowing about it is largely implicit (Vermersch, 1996) and an actor mixes what he actually does with what he should have done and what he could have done (Clot, 1999). On the other hand, attempts to account for "doing" through observation are insufficient because only the actor in question can make sense of her or his action (for more information, readers can consult Theureau, 2006). Finally, the third problem, and not the least significant, is how to account for "doing" in its collective dimension. Ergonomists and psychologists have developed individual approaches to "doing." They developed devices such as the "explicitation interview" (Vermersch, 1996), and so-called self-confrontation interview techniques mobilizing images of an actor in action, (e.g., the "simple" way [Theureau, 2006] or the "crossed" way [Clot, 2000]). However, these devices are designed to document the action of *one* player in the organization, and accounting for the collective is still a problem. Neverthless, the collective has been documented. For example, the collective perspective is found in Latour and Woolgar (1986), which documents two years in the life of the neuroendocrinology laboratory led by Pierre Guillemin at the San Diego Salk Institute. This investigation's objective was to document the life of the laboratory as a collective without addressing the inner experience of the actors involved. For a researcher documenting the collective perspective of organization, it is difficult to document also the subjective point of view of an individual actor involved in the same organization. It seems to be a difficult problem in the relationship between a researcher and the object of study: If one takes the individual point of view, it is then difficult to go back to the collective view; and if one places oneself at the collective level, it is hard to go to the individual level. A proposed solution is to use two researchers, one focuses on individuals and the other on the collective.

Extreme Situations of Management

Various authors have performed research in a variety of fields: polar and mountaineering expeditions, operation of special forces, fire rescue services, military operations, and a traditional Native American expedition, as well as innovative projects in the traditional sector of the economy. This research has led to many questions about varied issues that include the exploration stage of innovative projects in traditional companies, the connection between the expert practices of explorers and polar tourism, and the discussions and reflections by practitioners on flexibility in projects. Indeed, it seems research into polar expeditions projects and innovation projects are on the same page. They both deal with the management of extreme situations, and this concept has to be better clarified. An innovation economy based on knowledge has been emerging since the 1990s (Nonaka & Takeuchi, 1995; Amin & Cohendet, 2004; Foray, 2004), and this places at the heart of organizations the "design and implementation of collective action in the form of an innovative project, knowledge-intensive, in an evolutionary context, uncertain and risky" (Lièvre, 2014). Managers are faced with what we call, generically, "extreme situations management," which is to say management situations (Girin, 2011) immersed in a context that can be characterized as evolutionary, uncertain, and risky (Lièvre, 2005; Lièvre & Gautier, 2009). They are evolutionary because there is a break between a "before" and an "after" (Rivolier, 1998), which distinguishes phases or steps in the situation. They are uncertain because there is the possible emergence of the radically new (Orléan, 1986), which implies a "possible unpredictable" (Le Moigne, 1990). They are risky because the possibility of an undesirable event cannot be ruled out, and such an event could cause significant damage to an organization. These risks are not always measurable. The risk may also be the inability to reach the designated goal, and this represents a critical issue for an actor or collective.

It is also possible to make connections between these situations and research into reliability and organizational resilience as dimensions of organizational performance (Hoffnagel, Journé, & Laroche, 2009), as well as with research on organizational contexts qualified as "pluralistic" (Denis, Langley, & Rouleau, 2007).

A Complex Epistemological Position

Christophe Bredillet started the symposium, on which this book is based, by discussing complexity and mobilizing Le Moigne (1990) to build a framework capable of linking different paradigms of project management. We extend this thinking by addressing some epistemological issues related to the research in this book. The chapters take stock of past work, discuss contemporary results, and open up new and promising avenues of investigation. By recognizing the scientific nature of research's progressive nature (see Lakatos, 1976) through which further research is developed, it can be said that the research in this book is scientific. At the same time, this research is deliberately actuated (Avenier, 2010; Avenier, & Schmitt, 2007), that is, it wants to build knowledge for action by combining experiential knowledge and scientific knowledge (Lièvre, 2007). It differs from a conventional positivist posture and takes place in a new foundation of management science (David, Hatchuel, & Laufer, 2000; Martinet, 2006), which considers management science as a generic grammar of collective action with a leaning toward prescription (Simon) and whose object is organizing (Weick). It develops rooted organizational engineering (David, 2000) as part of a pragmatic, radical constructivist epistemology (Avenier, 2010). It is developed in the vein of intermediate theories (Glaser & Strauss, 1967) at the interface of the corpus of management science and the field of practice.

References

Amin, A., & Cohendet, P. (2004). *Architectures of knowledge: Firms, communities and competencies*. Oxford, UK: Oxford University Press.

Argyris, C. (1993). *Knowledge for action. A guide to overcoming barriers to organizational change*. San Francisco: Jossey-Bass.

Avenier, M. J. (2010). Shaping a constructivist view of organizational design science. *Organization Studies, 31*(9–10), 1229–1255.

Avenier, M. J., & Schmitt, C. (2007). *La construction de savoir pour l'action*. Paris: L'harmattan.

Blomquist, T., Hallgreen, M., Nilsson, A., & Söderholm, A. (2010). Project-as-practice: In search of project management research that matters. *Project Management Journal, 41*(1), 5–16.

Bonnet, E., & Lièvre, P. (2014). Logistique des situations extrêmes et management des connaissances: proposition d'un design de recherche dans le cadre du projet EuroMoonMars. *Management et Avenir,* no. 67, 224–242.

Clot, Y. (1999). *La fonction psychologique du travail*. Paris: PUF.

Clot, Y. (2000). Analyse psychologique du travail et singularité de l'action. In J.-M. Barbier et al., *L'analyse de la singularité de l'action* (pp. 53–70). Paris: PUF.

David, A. (2000). Logique, épistémologie, méthodologie en sciences de gestion: trois hypothèses revisitées. In A. David, A. Hatchuel, & R. Laufer, *Les nouvelles fondations des sciences de gestion* (pp. 83–108). Paris: Vuibert.

David, A., Hatchuel, A., & Laufer, R. (2000). *Les nouvelles fondations des sciences de gestion*. Paris: Vuibert.

Denis, J. L., Langley, A., & Rouleau, L. (2007). Strategizing in pluralistic contexts: Rethinking theoretical frames. *Human Relations, 60*(1), 179–215.

Foray, D. (2004). The economics of knowledge. Cambridge, MA: MIT Press.

Garel, G., & Lièvre, P. (2010). Polar expedition project and project management. *Project Management Journal, 41*(3), 21–31.

Gautier, A., Lièvre, P., & Rix, G. (2008). Les obstacles en matière d'apprentissage organisationnel au sein de l'organisation de la sécurité civile, une mise en perspective en termes de gestion des ressources humaines. *Politiques et Management Public, 26*(2), 137–168.

Girin, J. (2011). Empirical analysis of management situations: Elements of a theory and method. *European Management Review, 8*(4), 197–212.

Glaser, B. G., & Strauss, A. A. (1967). *The discovery of grounded theory*. Piscataway, NJ: Transaction Publishers.

Hatchuel, A. (2000). Vers une épistémologie de l'action collective. In A. David, A. Hatchuel, & R. Laufer (Eds.), *Les nouvelles fondations des sciences de gestion*. Paris: Vuibert.

Hoffnagel, E., Journé, B., & Laroche, H. (2009). Fiabilité et résilience comme dimensions de la performance organisationnelle: Introduction. *Management, 12*(4), 224–228.

Jarzabkowski, P. (2005). *Strategy as practice: An activity-based approach*. London: Sage.

Lakatos, I. (1976). Falsification and the methodology of scientific research programmes. In S. G. Harding (Ed.), *Can Theories be Refuted? Essays on the Duhem-Quine Thesis* (pp. 205–259). Springer Netherlands.

Latour, B., & Woolgar, S. (1986). *Laboratory life: The construction of scientific facts.* Princeton, NJ: Princeton University Press.

Le Moigne, J. L. (1990). *La modélisation des systèmes complexes.* Paris: Dunod.

Lièvre, P. (Ed.) (2001). *Logistique en milieux extrêmes.* Paris: Éditions Hermès.

Lièvre, P. (2005). *Vers une logistique des situations extrêmes.* HDR en Sciences de gestion, Université Aix-Marseille II, 256p.

Lièvre, P. (2007). La construction de savoirs pour l'action par intégration des connaissances pratiques tacites et de savoirs scientifiques classiques. In M. J. Avenier & C. Schmitt (Eds.), *La construction de savoirs pour l'action* (pp. 171–194), Paris: Editions L'Harmattan, Collection Action et Savoir.

Lièvre, P. (2012). *Du bon usage d'une clique en situation extrême: une expédition d'alpiniste en Patagonie. 4eme Journée de recherche AGRH-AIMS, "Management et reseaux sociaux,"* HEC Genève, 16 & 17 février.

Lièvre, P. (2014). Vers un management des situations extrêmes. *XXIII Colloque AIMS,* ESC Rennes School of Business, 26, 27, 28 mai.

Lièvre, P., & Gautier, A. (2009). Logistique des situations extrêmes: des expéditions polaires au service secours incendie. *Management et Avenir, 24,* 196–216.

Lièvre, P., Récopé, M., & Rix, G. (2003). Finalités des expéditeurs polaires et principes d'organisation. In P. Lièvre (Ed.), *La logistique des expéditions polaires à ski* (pp. 85–101). Paris: GNGL Productions.

Lièvre, P., & Rix-Lièvre, G. (2013). Leadership and organizational learning in extreme situations. Lessons of a comparative study of two polar expeditions: one of the greatest disasters (Franklin, 1855) and one the best achievements (Nansen, 1892). In C. M. Giannantonio and A. E. Hurley-Hanson (Eds.), *Extreme Leadership: Leaders, Teams and Situations Outside the Norm.* Northampton, MA: Edward Elgar.

Martinet, A.-C. (2006). Stratégie et pensée complexe. *Revue Française de Gestion,* no. 160, 31–45.

Midler, C. (1996). *L'auto qui n'existait pas.* Paris: Dunod.

Nonaka, I., & Takeuchi, H. (1995). *The knowledge creating company.* Oxford, UK: Oxford University Press.

Orléan, A. (1986). *Hétérodoxie et incertitude. Epistémologie et Autonomie, 5,* Les Cahiers du CREA, École Polytechnique, Paris.

Peretz, H. (2004). *Les méthodes en sociologie: l'observation.* Paris: Edition La découverte.

Rivolier, J. (1998). Stress et situations extrêmes. *Bulletin de Psychologie, 51*(6).

Rouleau, L., Allard-Poesi, F., & Warnier, V. (2007). Le management stratégique en pratiques. *Revue Française de Gestion,* no. 174, 15–24.

Rouleau, L., Chanal, V., Golsorkhi, R., & Langley, A. (2007). Regards croisés sur la perspective de la pratique en stratégie. *Revue Française de Gestion,* no. 174, 191–204.

Schatzki, T. R., Knorr-Cetina, K., & von Savigny, E. (2001). *The practice turn in contemporary theory.* London: Routledge.

Schön, D. (1983). *The reflexive practitioner: How professionals think in action.* New York: Basic Books.

Theureau, J. (2006). *Le cours d'action: méthode développée.* Toulouse: Octares.
Vermersch, P. (1996). *L'entretien d'explicitation.* Paris: ESF éditeur.
Vermersch, P. (1999). Introspection as practice. *Journal of Consciousness Studies, 6*(2–3), 17–42.
Whittington, R. (1996). Strategy as practice. *Long Range Planning, 29*(5), 731–735.

Afterword

Looking for the Ordinary in the Extraordinary!

Linda Rouleau

It was with great pleasure that I agreed to write this Afterword. When leafing through the manuscript of this book, I relived some powerful moments of the Darwin expedition in which I took part in autumn 2009 with my French colleagues Geneviève Musca and Marie Pérez, of Paris-Nanterre, and Yvonne Giordano, of the IAE de Nice. The expedition was subsidized by the French National Research Agency, and its goal was to follow a team of mountain climbers, "professional adventurers," as Grenier would say, who planned to traverse the Andean Cordillera, in Patagonia (Musca, Perez, Rouleau, & Giordano, 2009, 2010). Based on my experience that echoes several elements contained in this book, I share here some thoughts concerning the comprehensive and methodological contributions that polar expeditions and extreme situations can make to strengthen our understanding of project management.

Expeditions: Total Situations . . .

To strengthen our understanding of project management, there is interest in polar expeditions or any other undertaking in an extreme environment, because they are "total social situations," to use Goffman's term. Beyond their temporary nature, which they share with projects, these situations take place in a unique microcosm emerging through action and containing all the facets of a project, as well as of an organization or even of a social group. As this book demonstrates, we can read about the events on different levels, including a managerial perspective. We can seek to understand how leadership is exercised in situations of uncertainty and ambiguity (see Chapter 2). We can seek to understand the logistical, financial, strategic, and marketing challenges, as well as the risks run during these expeditions and experiences (see Chapter 8). And we can seek to understand even more than that, as these multifaceted entities can be examined through a variety of disciplinary lenses. Polar expeditions, just like projects, contain all the anthropological, cognitive, and symbolic phenomena that may help us to better understand group culture and behavior (see Chapter 6). In the same way, expeditions or extreme situations, like projects, can be analyzed from the angle of various theoretical corpora, including translation theory (see Chapter 1), network analysis (see Chapter 4), and the learning and construction of meaning (see Chapter 8).

Uniqueness makes the difference here. As Rix-Lièvre and Lièvre (see Chapter 5) said, expeditions and extreme situations allow us to see a "project as it is being made." They allow us to see in action the fundamental social processes that structure every human life, processes that management models often forget to take into consideration. In such contexts, key issues may stand out more clearly and the actors cannot hide behind routine activities and constructed discourses. The situation is new, and the identity of the group that exists only in the here-and-now of the expedition is in the process of being formed. Even though actors, according to Lehmann (see Chapter 9), always have the choice not to declare everything, they are forced in these situations to show their true selves, with their strengths, their weaknesses, and their contradictions. And if they do not do so at the necessary time, the group's survival is threatened. Thus, it is possible to identify the obscure processes of a fragile day-to-day experience, whose robustness and ordered dimension we often exaggerate.

. . . Anchored in the Socio-materiality of Social Ties

Expeditions and extreme situations require acknowledging the importance of two fundamental dimensions of action. First, equipment and tools occupy a central place, as they become both the protection and the extension of the body. In the expedition I was part of, material planning proved to be an important task to which the expedition leader and his followers gave an enormous amount of time and energy. In the expedition context, this materiality is radicalized in that it contrasts with the virginity of the destination while at the same time it symbolically creates a kind of shield against danger. Second, relationship skills, or to use Godé's expression (see Chapter 11), "the interrelated linkages of competencies," are also essential in these situations, as they are in project management. As several authors of this book clearly show, these linkages are put into action according to the actors' ability to draw on knowledge acquired in different contexts and to transfer it to unexpected situations.

In this sense, expeditions and extreme situations invite us to examine the socio-materiality of action (see Chapter 1). For example, during the Darwin expedition, the mountain, the sea, and the ship were actants because they were part of the unfolding action and made a difference in the course of events. One remarkable anecdote: The expedition members who were local actors gave these actants a personality when talking about them. One day the captain of our ship, noticing the wind picking up, said the following: "The mountain is telling us that it's seen enough of us and that we'd better leave right now." Expeditions and extreme situations also invite us to consider in our analyses the boundary objects that are part of this action. What the plan, the equipment, the mountain, the deliverables "make the actors do" is of interest for understanding how an expedition unfolded. In the same way, these boundary objects that are at the heart of a project should receive more attention from researchers. Outside this socio-materiality, the expedition leader—as well as the project manager—is lost.

. . . That Put Our Vision of Success into Perspective

A great lesson from analyzing these expeditions teaches that they never seem to unfold as planned, no matter how much effort is put into the planning stages. As Aubry and Lièvre remark, it is the same for project management: A significant number of the projects that are implemented in firms do not give the expected results. Yet they are not less successful because of that; it depends on how we define success. They often serve as experiments to implement successful future projects or to enable us to develop new competencies.

The Darwin expedition proved to be a series of micro-decisions that, for various reasons, had many unexpected consequences, which included not traversing the Darwin Cordillera as had been planned. When we look at the course of events in this expedition, as we would at those in more or less successful projects, one question immediately comes to mind: Why, when things are going badly, do they always get worse? It is certainly very difficult to answer this question. The political and cognitive dimensions that the literature typically uses to answer this type of question are inadequate. Moreover, at the start, actors agree to conventions and tacit rules that make them act and reproduce behavioral *patterns*. As the action unfolds, these *patterns* brand the actors, and they are difficult to break along the way. Thus, with each decision advancing the actors into the unknown, they continue to reproduce practices that they constructed during the day-to-day routine of the expedition or project. However, this agreement is tacit and unintentional, the result of multiple noncoordinated actions that various actors perpetuate in the action while maintaining a discourse of conquest and setting new targets. It would be interesting to see how these micro-dynamics form and re-form in projects that are failures or only partially successful.

. . . And Invite Us to Revise Our Research Practices

As fascinating as polar expeditions or extreme situations are for understanding project management, in the end, what contributes to the richness of collected data is the researcher's position and the data collection methods used. The scientific thinking in these areas is embryonic. However, in this book, Rix-Lièvre and Lièvre offer some insightful thoughts on these questions. For example, the researcher's dual position in fieldwork seems innovative to me. However, the fact remains that the "researcher-actor" is at the center of this type of process. Beyond the frameworks the researcher-actor puts in place to understand how the action is playing out based on actions and events, this process also requires from the researcher an active—and I would even say intense—emotional preparation. If there is an area where the emotional registers are complex and variable, it is this one. These emotional registers are at the very heart of the expedition, as each participant, whether an actor or a researcher, in the end engages his or her mind as much as his or her corporality. In my experience, a researcher's ability to collect and *a posteriori* analyze data is intrinsically linked to his or her ability not only to view the subject from a distance but also to view him or herself from a distance to be able to truly meet the Other.

In short, analyzing polar expeditions and extreme situations to better understand project management leads to a new school of thought. In the introduction to this book, Bredillet proposes nine schools of thought, each having their own specific ontological, epistemological, and theoretical perspectives. The book itself is proof that there is a very vibrant tenth school of thought, a "situated" or "practical" approach that invites researchers to think about the "fabric" of a project. There is no doubt that the analysis of expeditions and extreme situations enables us to see that these research areas provide a means for developing new theoretical ideas, for being creative in exploiting collected data, and for questioning the positioning of researchers who engage in seeking the ordinary through extraordinary experiences! Of course, to enter into unfolding collective action, whether of a polar expedition or a project, one must first and foremost agree to open oneself to different worlds, realities, and ideas. That is how we strengthen our knowledge. And that was just one lesson this expedition to the end of the world taught me!

References

Musca, G., Perez, M., Rouleau, L., & Giordano, Y. (2009). A practice view of strategic leadership in a highly risky and ambiguous environment: The Darwin expedition in Patagonia. *25th EGOS Colloquium*, Barcelona, July 2–4.

Musca, G., Perez, M., Rouleau, L., & Giordano, Y. (2010). Extreme organizational ethnography: The Darwin expedition in Patagonia. *26th Egos Conference*, Lisbon, July 1–3.

Index

A

adventure, 5, 23, 47, 75, 77, 78, 80, 87–97, 99–103, 105–109, 127, 200, 231
Afghanistan, 146–148, 153, 154
air-to-ground operations, 146
alternative scenarios, 116
alternative tourism, 87, 88
ambidexterity, 17–21, 24, 27, 29, 226, 231
ambiguity, 4, 149, 218
appraisal, 33, 34
automatism, 150

B

balance, 7, 40, 79, 101, 128, 205, 208, 209, 221, 225–227
behavior, 32, 33, 35, 61, 67, 68, 89, 135, 140, 141, 148, 152, 162, 175, 183, 186, 188, 190, 219, 230, 231
best practices, 214, 215
bibliographic accounts, 171, 172
budget, 47, 48, 116, 120, 123, 201–203, 205–208, 218
bundles of practices, 152
business, 54, 95, 99, 105, 106, 115–117, 119, 121–123, 176, 177, 213–215, 225

C

captain, 27, 151, 163, 226
Cheetah teams, 174, 186, 191, 192
closing, 21, 23, 24, 201, 204
cluster, 49
cognitive factor, 140, 141
collective action, 4, 59, 62, 69, 163
collective competence, 157–159, 162, 164–166
collective creativity, 150, 228
collective learning, 130, 136
collective memory, 85, 164, 165
collective roles, 219
combination of expertise, 160
combination of skills, 159, 161
commando, 160–163, 165, 167
commitment, 10, 11, 13, 21, 31–34, 36, 38, 40, 42, 64, 164, 165, 167, 174, 190, 213, 229
companion, 107, 129, 130
competency, 17, 19, 20, 28, 29, 45,

competency (*continued*) 47, 151, 153, 154, 176, 215, 225, 230, 231
complexity, 131, 135, 157, 219, 225, 227
constraints, 91, 139, 141, 206, 208, 211, 213, 215, 228
construction, 7, 14, 32, 50, 62, 67–69, 73, 91, 116, 130, 159, 162–164, 172, 174–176, 187, 192, 200, 202, 208
control, 11, 67, 93, 105, 106, 116, 119–123, 131, 137, 148, 163, 173, 177, 201, 203, 208, 214, 218, 220–222, 225–228, 230, 231
control systems, 116, 120–122
controversy, 4, 5, 7
cooperation, 46, 49–56, 77, 117, 121, 122, 124, 161
cooperative mechanism, 46, 52
coordination, 124, 138, 145–148, 150, 152–155, 163, 174, 190, 212, 228
creativity, 91, 150, 151, 153, 154, 186, 209, 218, 221, 222, 227, 228
Cree, 73–75, 77–79, 81, 85, 86
critical situation, 24–26, 220, 221, 223
cultural factor, 140, 141

D

danger, 11, 27, 64, 93, 94, 102, 130, 131, 163, 183, 187, 200, 202
debriefing, 3, 13, 38, 164, 165
decision making, 10, 64, 105, 116, 118, 121, 122, 164, 220, 222, 230
design stage, 3

E

ecological awareness, 87, 88
ecological tourism, 88
ecotourism, 88, 89, 91
environment, 10–13, 33–36, 38–41, 47, 61, 88, 89, 100, 103, 104, 106, 120, 129, 135, 145, 149, 153–155, 157, 175, 207, 208, 225, 227–230
ethnomethodology, 61
execution, 22, 79, 149, 160, 162, 164, 201–203, 206–209
expectations, 8, 10, 14, 24, 35, 104, 171, 175, 178, 186, 188, 189, 192, 213
expedition, 3–15, 17, 21–28, 31, 32, 36–42, 45–50, 53–55, 59, 60, 62–70, 73–75, 77, 79, 80, 85, 87–89, 91, 93, 95–97, 99–104, 106–109, 172, 176–186, 189, 190, 192, 199–204, 217–221, 223, 227, 229, 231
expedition travel, 87, 88, 97, 99–104, 108
experience, 4, 5, 7, 8, 10, 12–15, 22, 23, 31–36, 40, 47, 50, 55, 61, 63–70, 74, 75, 87–89, 93, 100, 101, 104–109, 129–131, 135–137, 140, 147, 150, 152–154, 157, 159, 162–164, 166, 167, 172, 200, 202, 213, 214, 223, 226, 229, 231
expertise, 9, 13, 64, 75, 101, 139, 148, 154, 160–162, 164, 167, 176, 192, 203, 219, 220, 226, 228–231
exploitation, 17, 19, 20, 22, 26–28
exploration, 3, 14, 17, 19, 20, 22, 26–29, 41, 88, 95, 96, 106, 108, 117, 118, 220, 225

extreme project teams, 157, 159
extreme situation, 4, 60, 135–138, 145, 146, 148, 150, 152, 153, 157, 159, 160, 164

F

feedback, 23, 121, 135, 136, 138–142, 153
financial guidance, 115, 120, 122
financial investment, 40
financial planning, 116, 123
financial sponsorship, 47
fire and rescue services, 136–141
fishing, 73–75, 77–80, 85
flexibility, 11, 74, 79, 80, 203, 205–209, 220–222, 225–227, 230, 231
French Special Forces (SF), 157, 160–162, 164–167
fundamental principles, 203, 217

G

generalist, 211, 214, 215
goal achievement, 171, 178, 186, 187, 192
golden triangle, 131
Greenland, 3–5, 7, 10, 12–14, 37
groundwork, 47, 48, 54

H

high dependence, 218
hit-or-miss, 218
hunting, 54, 75, 78, 79, 85
Hydro-Québec, 73, 74, 86

I

identity motivation, 103
impact assessment, 73, 74, 86
improvisation, 12, 15, 138, 163–165, 222
individual environment, 225
individual roles, 219
initiating, 52, 201
inspiration, 37, 231
Integrated Project Management Approach, 212
intentionality, 55, 56
interactive control systems, 120, 121
irrational move, 128

K

kayak, 3–8, 10–13, 106, 202
knowledge, 4, 5, 10, 12, 13, 15, 18, 19, 21, 22, 42, 47, 49, 61, 62, 65, 70, 74, 75, 85, 97, 118–120, 122–124, 128, 135, 137, 139, 140, 151–154, 165, 166, 191, 192, 213, 215, 223, 227, 230, 231
knowledge management, 4, 15, 135, 139, 215

L

laissez-faire, 226
language, 52, 54, 122, 136, 148–152, 163, 165
Latour, 3–5, 14, 45
Lazarus, 33, 34, 42
leader competency, 17
learning, 8, 11–13, 15, 17–20, 28, 49, 116, 121, 124, 130, 135, 136, 139–142, 146, 154, 166, 167, 192, 222, 225, 230
lessons learned, 127, 138, 204, 230
logic of the action, 63, 69

M

meaning, 21, 22, 28, 32, 35, 41, 60–62, 68, 95, 100, 139, 149, 150, 172, 230, 231
mental motivation, 102, 103
methodology, 14, 21, 59, 60, 66, 74, 138, 146, 157, 161, 163, 177, 215
metro, 205
military, 6, 95, 137, 138, 141, 145–147, 149, 154, 155, 220, 221
mission, 47, 103, 146–149, 153, 154, 160–167, 186, 187, 190, 211–213, 220
mobilization, 31, 33, 34, 36–38, 42, 49, 138, 162, 165
model, 6, 22, 89, 91, 105, 107, 120–124, 130, 141, 158, 159, 166, 167, 186, 212, 227, 229, 230
modes of action, 17, 20, 22, 27–29
monitoring and control, 201, 203
Montreal metro renovation project, 205
motivation, 31–37, 41, 42, 97, 101–103, 129, 140, 229
Mount Everest, 171, 172, 176, 177, 179, 184–189, 191
mountaineering, 48–50, 54, 130, 131, 185
Multimedia Logbook, 60, 63, 65
mutual trust, 15, 51, 150, 151, 153

N

necessity, 19, 75, 128, 171, 172, 178, 186, 188, 192, 229
network mobilization, 49
network theory, 46
norms, 31, 34–36, 41, 42, 140, 171, 172, 175, 186
Northern Québec, 73

O

observatory, 59, 60, 63, 68–70
observer-participant, 46
obstacle, 60, 81, 90, 141, 171, 178, 180, 186, 187, 192, 230
OPM3®. *See Organizational Project Management Maturity Model*
organizational enablers, 211, 212, 214, 215
organizational framework, 42, 138, 139, 141
organizational learning, 15, 17–19, 28, 135, 136, 139, 140, 166
Organizational Project Management Maturity Model (OPM3)®, 212
organizational view, 116, 124
overconcentration, 89
ownership, 139, 218

P

physical motivation, 101
planning, 11, 12, 18, 19, 21–23, 28, 48, 79, 116, 123, 127, 128, 130, 131, 173, 175, 177, 189, 199–208, 218, 222
planning risk, 199
polar expedition, 3, 4, 6, 9, 11, 12, 14, 15, 17, 21, 22, 31, 32, 36, 37, 42, 45–48, 53, 59, 60, 62–64, 67, 69, 70, 87, 199
polar regions, 6, 7, 87, 89
portfolio management, 212, 213, 215
postmission, 164, 166, 167
practices, 4, 12, 13, 18, 52, 59–63, 65–70, 99, 108, 121, 138–142, 145–148, 150, 152, 154, 155, 158, 171–176, 185–190, 192, 211, 212, 214, 215, 221, 230
pragmatism, 17

process groups, 201, 202, 204
profitability, 115, 122
program, 29, 60, 89, 205–208, 212, 215, 217
project, 3–15, 17–24, 26–29, 31, 33, 36, 37, 39–43, 45–55, 59, 60, 64, 65, 68–70, 73–80, 86, 95, 115–125, 127–131, 136, 157–161, 164, 165, 167, 171–177, 185–188, 190–192, 199–205, 207–209, 211–215, 217–222, 225–231
project environment, 11, 225, 227
project front-end, 116, 119, 120, 122–124
project leader, 12, 14, 15, 17, 19, 20, 22, 24, 26–29, 45, 225–231
project management, 3, 4, 8, 9, 11, 12, 14, 17, 18, 28, 29, 31, 45, 47, 49, 54, 64, 116, 118, 120, 124, 131, 157, 160, 165, 176, 190, 199–203, 211–215, 217, 218, 220–222, 225, 226, 229–231
project manager, 45, 131, 172, 176, 188, 192, 199, 200, 202–204, 207, 211, 213–215, 218–222, 227, 231
project mode, 158
project plan, 15, 18, 28, 117, 177
project planning, 127, 130, 201, 218
project team, 11, 45, 46, 51, 52, 55, 121, 157–159, 174, 176, 177, 191, 213

R

rationality, 20, 62, 173
rationality *in situ*, 62
reference systems, 163
reflexivity, 65
regulatory factor, 139, 141
reliability, 135, 205, 206
Reno-System, 205–208
rescue, 37, 88, 135–141, 171, 178, 179, 181, 182, 185, 186, 189, 190, 192, 202
rescue organizations, 135, 137
research and development, 217
resource allocation, 118, 207, 215
rigidity, 205, 206, 208, 209, 222, 226, 231
rigor, 18, 79, 165, 186
risk, 7, 12, 13, 26, 28, 50, 55, 61, 64, 70, 91, 93, 96, 97, 100, 103–108, 115, 116, 118, 120, 122–124, 130, 131, 135, 138, 141, 146, 154, 157, 165, 167, 173, 175–177, 181, 188, 189, 199–204, 207, 208, 214, 217, 218, 222, 223, 227, 228
risk assessment, 202
risk management, 70, 103, 135, 175–177, 188, 204, 208
risk society, 91
rules, 13, 91, 104, 137, 141, 146, 149, 151, 152, 165, 171–175, 178, 180, 186, 187, 189, 190, 192, 207, 221, 231

S

safety, 9, 14, 26, 37, 77, 88, 89, 91, 93, 99, 101, 105, 107, 108, 138, 206, 222
sensemaking, 137, 138, 141, 150, 153, 154, 166, 175
sensibility, 31, 35–37, 40–42
SF. *See* French Special Forces
shadowing, 147, 148
shared representation, 163, 165
Situated Practices Objectifying System (SPOS), 60, 65, 70
situated team, 171, 172, 185–190

social networks, 45–47, 49, 53, 55
socialization processes, 153–155
Société de transport de Montréal (STM), 205–208
sociological analysis, 46
socio-technical integration, 4
specialist, 13, 24, 26, 27, 47, 49, 124, 129, 131, 161, 163, 173, 211, 214, 215, 231
Spitsbergen, 47–50, 54
sponsors, 21, 24, 48, 54, 64, 74, 96, 129, 201–204, 213, 218, 221
sponsorship, 47, 215
sports climbing, 127
SPOS. *See* Situated Practices Objectifying System
standardized language, 148, 150, 152
STM. *See* Société de transport de Montréal
strong ties, 46, 49–51, 53–55
structural factor, 139, 141
subjective re situ interview, 66, 67
survival, 85, 96, 97, 151, 162, 171, 178, 179, 186, 188, 190, 192, 199

T

task, 11, 21, 39, 66, 79, 80, 117–119, 138, 139, 145, 148, 151, 153, 154, 158, 171, 173, 174, 176–179, 186–190, 192, 200, 202, 214, 219–222, 229
task-situated team, 186
team, 3–15, 21–28, 32, 37–43, 45–48, 51, 52, 55, 63–68, 74, 75, 78–80, 104, 117–119, 121, 122, 129–131, 136, 146, 150–155, 157–166, 171–178, 180–182, 185–192, 200, 202–204, 213, 214, 219, 225, 227–231
team environment, 225
team formation, 4, 8, 10, 14, 15, 171–174, 178, 180, 185, 186, 188, 191, 192
technology, 96, 97, 99, 105, 106, 108, 117, 122, 217
temporary organization, 21, 171–174, 176, 177, 185, 186, 191, 192
tiger team, 174, 190, 191
to read the path, 127–129
tools, 6, 9–11, 25, 60, 63, 64, 67, 69, 70, 115, 122, 135–137, 139, 140, 148, 152, 171, 172, 175, 176, 178, 186, 189, 191, 192, 208, 211, 213, 215
top management support, 213
traditional expedition, 97
traditional knowledge, 74, 85
trapping, 73–75, 78, 79, 85
trigger, 53, 108, 140, 171, 172, 178–180, 186–192, 199
turnover policy, 153–155

U

uncertainty, 14, 15, 93, 116, 118–121, 123, 129, 146, 157, 166, 228, 230, 231
unexpected, 40–42, 88, 106, 150, 151, 154, 162, 171, 173, 191, 227
unknown, 18, 91, 93, 95, 129, 217, 219–222

V

values, 11, 15, 23, 33–35, 42, 49–51, 54, 55, 61, 74, 83, 85, 86, 89, 97, 100, 104–106, 117, 118, 120–124, 140, 141, 160, 165, 171, 172, 174, 175, 177, 192, 231
verbalization, 62

W

weak ties, 46, 49–55
Weick, 11, 15, 59, 129, 136, 137, 151, 165, 166, 171, 172, 174, 175, 178, 186, 187, 189